Programming with MATLAB for Scientists

A Beginner's Introduction

Programming with MATLAB for Scientists

A Beginner's Introduction

Eugeniy E. Mikhailov

CRC Press
Taylor & Francis Group
Boca Raton London New York

CRC Press is an imprint of the
Taylor & Francis Group, an **informa** business

Published in 2017 by CRC Press
Taylor & Francis Group
6000 Broken Sound Parkway NW, Suite 300
Boca Raton, FL 33487-2742

CRC Press is an imprint of Taylor & Francis Group

No claim to original U.S. Government works

Printed in the United States of America on acid-free paper

10 9 8 7 6 5 4 3 2 1

International Standard Book Number-13: 978-1-4987-3828-6 (paperback)

Library of Congress Cataloging-in-Publication Data

Names: Mikhailov, Eugeniy E., 1975- author.
Title: Programming with MATLAB for scientists : a beginner's introduction / Eugeniy E. Mikhailov.
Description: Boca Raton, FL : CRC Press, Taylor & Francis Group, [2017]
Identifiers: LCCN 2017031570 | ISBN 9781498738286 | ISBN 1498738281
Subjects: LCSH: MATLAB. | Science–Data processing. | Engineering mathematics–Data processing.
| Numerical analysis–Data processing.
Classification: LCC Q183.9.M55 2017 | DDC 510.285/53–dc23
LC record available at https://lccn.loc.gov/2017031570

Visit the Taylor & Francis Web site at
http://www.taylorandfrancis.com

and the CRC Press Web site at
http://www.crcpress.com

Contents

Preface

Intended Audience

This book is intended for anyone who wants to learn how to program with MATLAB and seeks a concise and accessible introduction to programming, MATLAB, and numerical methods. The author hopes that readers will find here all necessary materials for handling their everyday computational and programming problems. Even more seasoned readers may find useful insights on familiar methods or explanations for puzzling issues they might encounter.

We will start with simple concepts and build up a skill set suitable to model, simulate, and analyze real-life systems. Additionally, this book provides a broad overview of the numerical methods necessary for successful scientific or engineering work. We will get familiar with a "lore" of computing, so you will know what to look for when you decide to move to more advanced techniques.

The book is based on material of the one semester "Practical Computing for Scientists" class taught at the College of William & Mary for students who have not yet declared a major or students majoring in physics, neuroscience, biology, computer science, applied math and statistics, or chemistry. The students who successfully took this class were at all levels of their academic careers; some were freshmen, some where seniors, and some were somewhere in between.

Why MATLAB?

A couple words about MATLAB, as it is our programming language of choice. MATLAB has a good balance of already implemented features, which are important for scientists and for ease of learning. MATLAB hides a lot of low-level details from users: you do not need to think about variable types, compilation processes, and so on. It just works. You can also do a calculation on a whole array of data without tracking every element of the array. This part is deep inside of MATLAB.

From an instructor's point of view, you do not need to worry about the installation of MATLAB for your class. It is easy and students are capable of doing it alone. More importantly, it looks and works the same on variety of operational systems, such as Windows, Mac, and Linux. MATLAB produces exactly the same result on all computers.

From a student's point of view, MATLAB is probably the most frequently required programming language for an engineering or scientific position. Therefore, if you learn MATLAB now, you likely will not need to retrain yourself to another industry standard programming language.

MATLAB has a downside: it is expensive to purchase if your school or workplace does not provide it. This is not a big worry; you can do exercises from all the chapters, except the data fitting, with a free alternative: GNU Octave. The fitting in Octave uses a different set of commands, but everything else will work the same (you might need minor tweaking for more advanced options).

What Is not Covered in This Book?

This book does not extensively cover MATLAB's commands. There is no reason to write another manual for MATLAB (which has an excellent one already) or redo tutorials available on the web.

This book is also not a substitute for a book explaining the ins and outs of numerical methods. Whenever possible, we discuss what interesting things can be done with a numerical method and do not bother with the most efficient implementation. However, the beginning of the book is an exception. There, the basics of programming are explained via implementations of some numerical method algorithms (which often have MATLAB's built-in equivalents already).

How to Read This Book

If you have programmed before, you can skip the majority of Part I, but make sure that you are fluent with the element operations described there, the differences between array and element-wise operations, and array slicing.

If you are scientist, then the plotting and fitting materials are a must. Make sure that you have read the fitting chapter. If you need to learn anything important about data analysis, then learn this.

The material in Part III is somewhat optional, although the author strongly recommends the optimization problem chapter (see Chapter 13). It is amazing to see how many other problems are essentially optimization problems and can be solved with methods presented there. The time to use the material in this section will probably come in upper undergraduate classes.

As you get more and more fluent with programming, reread the good programming practice materials in Section 4.4 and try to implement more and more techniques from there.

Data files and MATLAB listings locations

All MATLAB's listing and required data files used in the book are available on the web.* The PDF version of this book contains direct links to such files.

MATLAB® is a registered trademark of The MathWorks, Inc.
For product information, please contact:
The MathWorks, Inc.
3 Apple Hill Drive Natick,
MA 01760-2098 USA
Tel: 508 647 7000 Fax: 508-647-7001
E-mail: info@mathworks.com
Web: www.mathworks.com

---

* Please visit http://physics.wm.edu/programming_with_MATLAB_book.

Part I

Computing Essentials

CHAPTER 1

Computers and Programming Languages: An Introduction

This chapter defines programming and programming languages for the reader with a summary of modern computing as a backdrop. It goes further to explain the role of numbers in computers and potential problems in their use.

> Computers are incredibly fast, accurate, and **stupid**. Humans beings are incredibly slow, inaccurate, and brilliant. Together they are powerful beyond imagination.
>
> **Leo Cherne**[*]
> *(1969)*

1.1 Early History of Computing

It is hard to believe that only about 100 years ago, there were no computers. Computers used to be humans who were trained to do quite complex calculations using only their minds and some mechanical aids. Have a look at Figure 1.1. The painting[†] depicts an ordinary school in rural Russia. It might be hard to see, but pupils are looking for the result of the following expression inscribed on the blackboard:

$$\frac{10^2 + 11^2 + 12^2 + 13^2 + 14^2}{365} \tag{1.1}$$

There were aids: abaci, sliding rulers, pre-calculated tables of functions (logarithms, trigonometric functions, exponents, ...), and mechanical calculators, which start to show up at the end of the nineteenth century. These mechanical aids were not programmable; they were designed to do a fixed subset of elementary calculations. Instead, you would "program", that is, ask to do a complex problem, a human.[‡]

---

[*] The true origin of this quote is not 100% certain.

[†] The painting is in the public domain and was obtained from www.wikiart.org/en/ nikolay-bogdanov-belsky/mental-arithmetic-in-the-public-school-of-s-rachinsky.

[‡] We can argue that this was the zenith of programmability. You could just specify what to do in a normal human language and tasks would be done. Nowadays, we have to split the task into excruciatingly small elementary subtasks so that a computer can understand.

Figure 1.1   On the left is the painting by Nikolay Bogdanov-Belsky, dated 1895: "Mental Arithmetic. In the Public School of S.Rachinsky". An enlarged portion of the blackboard is shown on the right.

1.2   Modern Computers

We can probably attribute the title of the very first computer, the direct ancestor of modern ones, to ENIAC (Electronic Numerical Integrator And Computer), which was constructed in 1946. Here are some of its specifications:

- Weight: 30 tons

- Cost: $500,000 ($6,000,000 adjusted for inflation)

- Power consumption: 150 kW (averaged consumption of 500 households)

With all its might, ENIAC could do the following in 1 s: 5000 additions, 357 multiplications, or 38 divisions. Modern computer speed is measured in FLOPS (the number of floating-point operations per second). So we see that ENIAC was able to do about 100 FLOPS, while the author's several-years-old desktop computer can do about 50 **Mega** FLOPS. A typical mobile phone outperforms ENIAC by many orders of magnitude.

1.2.1   Common features of a modern computer

A modern computer typically has the following features. It has one or more central processing units (CPU), memory, which holds data and programs, and input and output interfaces (keyboards, hard drives, displays, printers, ...). A typical computer uses the binary system internally.* Despite these differences, the main

---

* There were attempts to use ternary system computers, that is, based on the number 3. This has several benefits, but it is more costly in hardware.

feature that separates computers from their lesser ancestors is that computers can be programmed for any general task without changing their hardware.

1.3   What Is Programming?

If computers are programmable, then we should be able to program them, that is, generate a list of instructions suitable for execution by a computer. "Generate a list of instructions" does not sound scary. What is the problem here? This is not a trivial task; in fact, some refer to it as the art of programming.

Think about the following situation: you want to eat. So, the simplest program to yourself: buy a pizza and eat. Sounds easy. Now, you start to split it into the list of instructions suitable to your body: pick up a phone (mind that you need to find it first, which is not trivial), punch numbers (imagine how many instructions are needed for your arm just to flex a finger, point to a number, and punch), talk to a human on the other end of the phone and make an order (this not trivial at all if you try to split it into elementary operations), wait for delivery, pay for delivery (open a door, talk, find a wallet, count money, give money, get change), bring a box to the kitchen, open the box, get a slice, eat. Each of these operations needs to be split into a set of even more elementary ones.

When you program a computer, not only do you need to implement all of these tiny details, but you also have to do it in a foreign (to you) language, which is not even designed for humans.

However, the author put all of the above into the "coding" category. This is relatively easy; a harder problem is to think ahead and design a safety net for every operation (what to do if the pizza is cold or you dial a wrong number). But this is still relatively easy. The ability to spot mistakes in seemingly correct program flow and fix it (i.e., *debugging*) is the difficult skill to acquire. Scientific programming is even harder; you are often the first person doing a particular calculation, and there is no reference to check it against. For example, how would you know that your calculation of the trillionth digit of π is correct if you are the first to do it?

Whenever possible, the author will show alternative routes to check or test programs throughout the book. Eventually, you should look for test cases on a subconscious level when you program. Anyone can code, but only a true master can say with some certainty[*] that the program is correct and does what it was designed for.

1.4   Programming Languages Overview

There are literally hundreds of programming languages. Unfortunately, none of them is the "silver bullet" language that is good for every situation.[†] Some of the programming languages are better for speed, some for conciseness, some

---

[*] Computer scientists have a theorem that states that it is impossible to prove the correctness of a general (complex enough) program. It is the so-called *halting problem*.

[†] The author knows only 10 of all languages and actively uses about 4 in everyday work.

for controlling hardware, some for reducing mistakes, some for graphics and sound output, some for numerical calculations, and so on. The good news is that majority of the languages are somewhat similar. The situation resembles human languages: once you know that there are nouns, verbs, and adjectives, you find them in all languages, and all you need to know is the proper grammar for a given language.

There are several ways to categorize programming languages. One is how low or high level they are. In our example with the pizza order, the high-level instruction set would be order and eat; the low-level program would be very detailed, including the timing of every nerve pulse to flex your arms and fingers.

The lowest-level programming language is undoubtedly the binary code. This is the only language that computers understand, and every other programming language is translated into this one when the program is executed by a computer. The binary code is not for humans, and no humans use it these days, except maybe people working with memory-starved microprocessors. A few low-level languages are assembler, C, C++, Forth, and LISP. A few high-level languages are Fortran, Tcl, Java, JavaScript, PHP, Perl, Python, and MATLAB®. The divide between low and high levels is in the eye of the beholder. MATLAB, which is very concise when you work with numerical calculations, will become extremely wordy if, for example, you try to find a file with the word "victory" in the name among all your files.

Another way to categorize a programming language is the internal implementation. With certain languages, a program needs to be *compiled*, that is, translated into binary code, as a whole before execution. With other languages, the commands are *interpreted* one by one as they arrive at the execution queue.

A good example highlighting the differences between two approaches would be the following. Imagine someone gave you a paper in a foreign language and asked you to read it in front of people. You can translate it ahead of time, and just read the fully translated (compiled) version. Alternatively, you can do it at the time of the presentation by translating it line by line (i.e., interpreting it). With this example, we can see advantages and disadvantages of each approach. If you need speed during the presentation, do the full translation or compilation; then, you will have extra capacity to do something else during the presentation. But this takes time for preparation. You can interpret the paper on the fly, but than you will be loaded with this task and will not be able to do side activities. However, if you are working with a paper in progress, where there is back and forth between the intended audience and the author, it would be painful to translate the paper fully from the beginning every time (unlike humans, computers have no recall ability and do everything from scratch). Debugging, which is described above, is much easier to do interactively: find a problem spot, fix it, and continue. There is no need to redo the beginning, since it is the same. In this situation, the interpreted programming languages really shine.

There is also a third, somewhat in-between, category, where the program is pre-translated into something that is not yet binary code, but a much simpler and easier-to-interpret language. But from our point of view, this is similar to the languages that need to be compiled, since we cannot interactively debug our programs with this approach.

MATLAB in this regard is an interactive language,* so it is fun to work with it. Think of a command, run it, observe results. If the command has a mistake (*bug*), you can fix it without redoing hours of preliminary calculations.

> ### Words of wisdom
>
> Computers do what you ask, not what you wish. If a computer delivers an unsatisfactory result, it is most likely because you did not properly translate your wish.

1.5 Numbers Representation in Computers and Its Potential Problems

1.5.1 Discretization—the main weakness of computers

Let's have a look at the following expression:

$$1/6 = 0.1666666666666666 \cdots$$

Because the row of 6s goes to infinity, it is impossible to keep this number in a computer with infinite precision. After all, the computer has a finite memory size. To circumvent the demand for infinite memory, the computer truncates every number to a specified number of significant digits.

For example, let's say it can hold only four significant digits. So,

$$1/6 = 0.1667_c$$

Here, the subscript "c" stands for computer representations. Notice that the last digit is not 6 anymore; the computer rounded it to the nearest number. This is called *round-off error* due to truncation or rounding.

Due to the rounding, all the following numbers are the same from the computer's point of view:

$$1/6 = 1/5.999 = \mathbf{0.1667}123 = \mathbf{0.1667}321 = \mathbf{0.1667}222 = \mathbf{0.1667}$$

We might arrive at the paradoxical (from the algebraic point of view) results:

$$20 \times (1/6) - 20/6 \neq 0$$

Since parentheses set the order of calculations, the above is equivalent to

$$20 \times (1/6) - 20/6 = 20 \times 0.1667 - 3.333 = 3.334 - 3.333 = 10^{-4}$$

---

* If you really need the speed, you can compile the mission-critical portions of your code, though this is beyond the scope of this book.

Even if we allow more digits to store a number, we will face the same problem but at a different level of precision.

1.5.2   Binary representation

Now, let's talk a bit about the internals of a modern general-purpose computer. Deep inside, it uses a so-called *binary* system. The smallest unit of information, called a *bit*, can have only two states: yes or no, true or false, 0 or 1. Here, we stick to the 0 and 1 notation. A bit is too small a unit. A larger unit is a *byte*, which is collection of 8 bits. The byte can represent $2^8 = 256$ different states, which can encode numbers from 0 to 255 or $-128 \ldots 0 \ldots 127$ or a symbol of an alphabet.

The days of 8-bit computers are over, and a typical chunk of information consists of 8 bytes, or 64 bits. Consequently, 64 bits can encode integer numbers in the range[*]

$$-2,147,483,648 \cdots 0 \cdots 2,147,483,647$$

The range looks quite large, but what happens when we ask the computer to calculate $2,147,483,647 + 10$? Surprisingly, the answer is $2,147,483,647$. Note that we see 47 at the end, and not the expected 57. This is called the *overflow error*. The situation is equivalent to the following. Imagine that someone can count using only the fingers of his hands, so the largest number in this system is 10.[†] If someone asks him to add 2 to 10, the person will run out of fingers and will report back the largest available number, which is 10.

1.5.3   Floating-point number representation

Numbers that have a decimal point are called *floating-point* numbers, for example, 2.443 or 31.2×10^3. Let's see what a computer does when it encounters a floating-point number, for example, a negative number such as -123.765×10^{12}. First, the computer converts it to the scientific notation with only one significant digit before the point:

$$(-1)^{s_m} \times m \times b^{(-1)^{s_e} q} \tag{1.2}$$

where:
   s_m is the sign bit of the mantissa (1 in our case)
   m is the mantissa (1.23765)
   b is the base of the exponent (10)
   s_e is the sign bit of the exponent (0 in our case)
   q is the exponent (14).

---

[*] If the situation changes in the future, you can check the range in MATLAB with the intmin and intmax commands.
[†] Actually, it is $2^{10} = 1024$ if we use the binary notation.

There is a caveat: the computer transforms everything into the binary system, so $b = 2$. Also, we recall that we have only 64 bits to store sign bits, mantissas, and exponents. According to the Institute of Electrical and Electronics Engineers (IEEE) 754 standard, mantissas take 52 bits plus 1 bit for the sign (this is equivalent to about 17 decimal digits). The exponent takes 10 bits plus 1 sign bit (this is equivalent to roughly $10^{\pm 308}$).

This representation has the following limitations. The largest positive number[*] is $1.797693134862316 \times 10^{308}$. The smallest positive number[†] is $2.225073858507201 \times 10^{-308}$.

Consequently, the following:

$$(1.797693134862316 \times 10^{308}) \times 10 = \infty \tag{1.3}$$

produces *overflow error*. At the same time,

$$(2.225073858507201 \times 10^{-308})/10 = 0 \tag{1.4}$$

produces *underflow error*.[‡] Finally, we show two examples of the *truncation error*:

$$1.797693134862316 + 20 = 21.797693134862318$$
$$1.797693134862316 + 100 = 101.7976931348623\_\_$$

Notice what happened with the bold numbers in these examples.

> **Words of wisdom**
>
> Computers are never exact with numerical calculations. At the very best, they are accurate to a certain precision, which is often worse than a theoretically achievable single number.

How to mitigate the above situations? Use numbers of a similar magnitude, and do not rely on the least significant digits in your answers.

1.5.4 Conclusion

Despite the quotation at the beginning of this chapter, we just saw that computers are not that accurate. But at any rate, **computers are not a substitute for a brain**. The answers produced by computers should always be checked. This book will provide us with the means to complement the human ability to think with the aid of computers.

---

[*] Use realmax in MATLAB.

[†] Use realmin in MATLAB.

[‡] On some computers, you might get a meaningful result ($2.225073858507201 \times 10^{-309}$) due to some representation trickery. But $(2.225073858507201 \times 10^{-308})/10^{17}$ is guaranteed to fail. Why this is so is beyond this scope of this book. It is explained in the IEEE Standard for Floating-Point Arithmetic (IEEE 754).

1.6   Self-Study

Prerequisites: If you are new to MATLAB, please read Chapter 2 first.

Problem 1.1
Find the largest number x (one significant digit is enough) such that the numerical evaluation of the expression

$$(1+x) - 1$$

equals to zero. The value of x gives you an estimate of the relative uncertainty of your calculations with MATLAB; try to keep it in mind when you do calculations. Note that x is actually rather small.

Problem 1.2
Find the value of the expression

$$20/3 - 20 \times (1/3)$$

Algebraically, you should get zero. If your result is not zero, please explain.

Problem 1.3
Find the numerical value of the expression

$$10^{16} + 1 - 10^{16}$$

with MATLAB. Algebraically, you should get 1. If your result is not 1, please explain.

Problem 1.4
The base of natural logarithms can be expressed as

$$e = \lim_{n \to \infty} \left(1 + \frac{1}{n}\right)^n \qquad (1.5)$$

so the numerical estimate for e should improve at larger and larger n. Find the n value (to the order of magnitude) at which the numerical errors lead to drastic deviations from the true $e = 2.718281828459\ldots.$

CHAPTER 2

MATLAB Basics

This chapter provides an introduction to the basics of MATLAB's machinery. It describes MATLAB's graphical user interface, the use of MATLAB as a powerful calculator, and key functions and operators, such as the range operator. It explains how to edit MATLAB code efficiently, and the use of matrices and plotting in MATLAB.

By no means is this chapter intended to replace MATLAB's comprehensive documentation. This very short introduction is only the basics of MATLAB's machinery. The reader is strongly urged to read the relevant documentation once the material is introduced. Many examples in this book show only subsets of the capabilities for given commands or functions.

2.1 MATLAB's Graphical User Interface

When you start MATLAB, the graphical user interface (GUI) resembles the view depicted in Figure 2.1. The GUI consists of several sections: action menus at the very top, view of the file system in the sub window labeled "Current Folder," preview of a selected file labeled "Details" in the left lower corner, "Workspace" at the right, and, most importantly, the window labeled "Command Window," which is situated in the middle.

If you are new to MATLAB, it is a very good idea to click on the "Getting Started" link at the top of the command window. This will redirect you to MATLAB's documentation with several tutorials to help you to get started.

The "Command Window" is where you can type your commands to MATLAB and see the results of their execution. At the very least, MATLAB can be used as a very powerful calculator. If we type 2 + 2 and hit the <enter> key, the MATLAB window will look similar to Figure 2.2. Note that we have obtained our requested answer: 4; it is assigned to a *special variable* ans, which is short for "answer." This variable always has the result of the last *unassigned* MATLAB command evaluation. If you look at the right side of Figure 2.2, you will notice that the "Workspace" window has changed. It shows that the variable ans exists and its value is 4. This will be true for all variables that we define during our execution.

To avoid the use of screenshots, we will use transcripts of the computational sessions. The above 2 + 2 calculation can be shown as

```
>> 2+2
ans =
     4
```

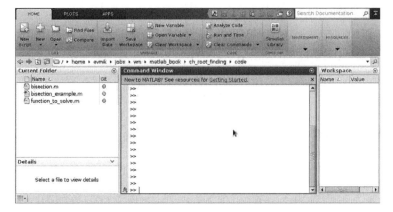

Figure 2.1   MATLAB window at start.

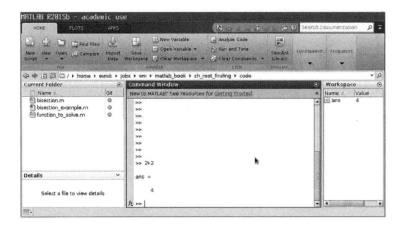

Figure 2.2   MATLAB window after $2 + 2$ calculation.

The lines marked with >> depict our commands, and everything else is the result of the commands, execution.

The ans variable can be used in calculations. For example,

```
>> ans*10
ans =
    40
```

produces 40, since ans used to be 4, per the result of the previous calculation. If we continue to do calculations, ans will automatically update its value:

```
>> ans+3
ans =
    43
```

As you can see, MATLAB uses the most recent value of the ans variable assigned during previous calculations.

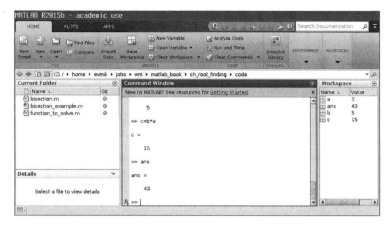

Figure 2.3   MATLAB window after assignments of the variables "a," "b," and "c."

We can define other variables and use them in our calculations:

```
>> a=3
a =
        3
>> b=2+a
b =
        5
>> c=b*a
c =
       15
```

We have assigned the results of the execution of the above commands to variables a, b, and c. The MATLAB window now looks like Figure 2.3. The "Workspace" now shows the newly assigned variables and their values, as you can see in Figure 2.3. Notice the apparently strange value of ans. It is 43, as it was set a while ago with the ans+3 command. Since then, we did not make any **unassigned** calculation, that is, all results of following calculations were assigned to the corresponding variables.

Words of wisdom

Avoid the unnecessary use of the ans variable in your calculation. Its value might change after a command execution. It is much better to assign the calculated result to a named variable. In this case, you are in control.

Often, there is no need to examine the result of intermediate calculations. In the previous example, we likely only needed to see the result of the very last expression c = b * a. You can end an expression with ; to *screen out* or *suppress the output* of an expression evaluation. Compare the following transcript with the previous one:

```
>> a=3;
>> b=2+a;
>> c=b*a
c =
      15
```

2.2　MATLAB as a Powerful Calculator

2.2.1　MATLAB's variable types

MATLAB allows you to have variables with arbitrary names with letters, digits, and underscores. You can mix in arbitrary order with only one requirement: the variable name must not start with a digit. Examples of valid names are a, width_of_the_box, test1, and medicalTrial_4. The variable name is just a label for the content stored in it. We will be mostly concerned with variables of the following numerical types.

Integer numbers

- 123
- −345
- 0

Real numbers

- 12.2344
- 5.445454
- MATLAB uses the engineering notation, so 4.23e−9 is equivalent to 4.23×10^{-9}
- pi is the built-in constant for $\pi = 3.14159265358979323846264338327950 2 \ldots$

Complex numbers

- 1i is equivalent to $\sqrt{-1}$
- 34.23+21.21i
- (1+1i)*(1−1i) = 2

We also need to be aware of variables of the *string* type (see also Section 2.5.3). They are typically used for labels, messages, and file names. To make a string, surround your words with apostrophes.*

Strings

- `'programming is fun'`

- `s='debugging is hard'`

This is not a complete description of variable types, but it is good enough to start using MATLAB productively.†

2.2.2   Some built-in functions and operators

MATLAB has hundreds of built-in functions and operators. Here, we cover only the basic ones needed in everyday life. If you have ever used a calculator, then the material in this section should be familiar to you.

- Trigonometry functions and their inverses, which are in **radians** by default
  - `sin, cos, tan, cot`
  - `asin, acos, atan, acot`

```
>> sin(pi/2)
ans =
     1
```

There are also trigonometry functions that are in degrees by default
- `sind, cosd, tand, cotd`
- `asind, acosd, atand, acotd`

```
>> sin(90)
ans =
     1
```

- Hyperbolic functions
  - `sinh, cosh, tanh, coth`
  - `asinh, acosh, atanh, acoth`

---

* As a historical artifact, MATLAB does not follow the typographical convention by which you need to use opening and closing quotation marks.

† There are numerical types that specify the internal representation of a number to a computer: `uint`, `int32`, `single`, `double`, and so on. These are often needed when you communicate with hardware. There is also a way to store a reference to a function, that is, a so-called *handle* (see Section 4.5.1).

- Logarithms
  - log for the natural logarithm
  - log10 for the logarithm with base of 10
- The exponentiation operator and function
  - for x^y use x^y or alternatively power(x,y)
  - for e^y use exp(y)

2.2.2.1   Assignment operator

The *assignment* operator is depicted as =. As its name implies, it assigns[*] the value of the right-hand expression of = to the left-hand variable name.

The MATLAB expression x = 1.2 + 3.4 should be read as

- Evaluate the expression at the right-hand side (RHS) of the assignment operator (=)
- Assign the result of the RHS to the variable on the left-hand side
- Now the variable with the name x holds the value 4.6

We are free to use the **value** of the variable x in any further expressions.

```
>> x+4.2
ans =
      8.8000
```

2.2.3   Operator precedence

Look at the following MATLAB expression and try to guess the answer: −2^4*5 + tan(pi/8+pi/8)^2. This could be quite hard. We might be unsure as to what order MATLAB does the calculation. Does it first calculate tan(pi/8+pi/8) and then square it; or does it first calculate (pi/8+pi/8)^2 and then take tan?

This is all controlled by MATLAB's *operators precedence* rules. Luckily, it follows standard algebraic rules: the expressions inside the parentheses are calculated first, then functions evaluate their arguments, then the power operator does its job, then multiplication and division, than addition and subtraction, and so on.

So, our example will be simplified by MATLAB during the expression evaluation as

1. −(2^4)*5 + (tan((pi/8+pi/8)))^2
2. −(16)*5 + (tan(pi/4))^2
3. −80 + (1)^2 = −80 + 1= −79

The final result is −79.

---

[*] People often confuse it with the *equality check* operator. We will see the important difference between them later, in Section 3.4.

To see the full and up-to-date rule set, search for the word precedence in the help browser or execute doc precedence in the command window.

> **Words of wisdom**
>
> If you are not sure about operation precedence, use parentheses () to enforce your way.

2.2.4 Comments

MATLAB treats everything after % as a *comment*. It helps a reader to understand what is going on.

```
>> % this is the line with no executable statements
>> x=2 % assigning x to its initial value
x = 2
```

The comment has no influence on the command execution.

```
>> y=4 % y=65
y = 4
```

Note that y is assigned to be 4 and not 65.

2.3 Efficient Editing

Some of us have learned touch typing and type quickly. Others, such as the author, type slowly. In either case, MATLAB has some built-in capabilities to help us type and edit more efficiently.

The *completion* feature is one of the most useful. When you are entering a function or a variable name, type a couple of the first symbols from its name and then hit the <tab> key. You will get

- Either a fully typed name (if it is unique)
- Or a little chart with choices for the possible name completions
  - Use <up> and <down> arrows to highlight intended choice
  - Alternatively, use <Ctrl-p> and <Ctrl-n>
  - Then hit <enter> to finalize your choice

This works in the command line as well as in MATLAB's editor.

For example, if you type plot in the command window and hit the <tab> key, MATLAB will show you all possibilities that start with "plot," as shown in Figure 2.4.

Sometimes, you may need to refresh your memory on the order of arguments that are used by a function. Just type the function name with the opening parenthesis and wait a bit. MATLAB will show you the intended way to use this function.

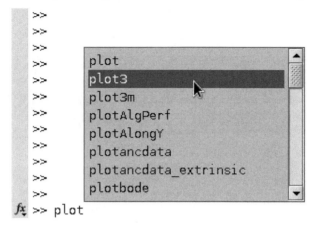

Figure 2.4   MATLAB's function completion example.

```
>>
>>
>>
>>
plot(X,Y)
>>
plot(X,Y,LineSpec)
>>
plot(X1,Y1,...,Xn,Yn)
>>
plot(X1,Y1,LineSpec1,...,Xn,Yn,LineSpecn)
>>
plot(Y)
>>
plot(Y,LineSpec)
>>
plot(___,Name,Value)
>>
plot(ax,___)
>>
plot(___)
>>
                                    More Help...
fx >> plot(
```

Figure 2.5   MATLAB's function arguments hints example.

For example, if you type plot(and pause, the relevant part of the MATLAB's window will look like Figure 2.5.

2.4   Using Documentation

MATLAB has excellent documentation. If you are unsure about something, consult MATLAB's help files. You can access them via the help drop down menu. Alternatively, you can call for the documentation directly in the command window:

- docsearch word
  - Search for the word in the help files and show help files where this word is present.
  - Example: docsearch trigonometry
- help name
  - Outputs a short help text directly into the command window about function or method named name.

- Example: `help sin`

- `doc name`
  - Shows a reference page about the function/method named name in the help browser. Usually, `doc name` produces more information in comparison to `help name`.

  - Example: `doc sin`

2.5 Matrices

MATLAB can work with matrices. If you are new to matrices, they are nothing more than tables of elements. Scientists use the words "matrix," "array," and "table" interchangeably.

2.5.1 Creating and accessing matrix elements

Let's create a 3×5 matrix (3 rows and 5 columns).

```
>> Mz=zeros(3,5)

Mz =

     0     0     0     0     0
     0     0     0     0     0
     0     0     0     0     0
```

This is not the only way, but it is the one that ensures the matrix is filled with zeros. Note that the first argument of the function `zeros` is the number of rows, and the second one is the number of columns.

Note: it is possible to have more than two-dimensional matrices, but they are difficult to visualize. For example, try `zeros(2,3,4)`.

We can access and assign an arbitrary element of a matrix. Let's set the element in row 2, column 4 to 1.

```
>> Mz(2,4)=1   % 2nd row, 4th column

Mz =

     0     0     0     0     0
     0     0     0     1     0
     0     0     0     0     0
```

Note that such an assignment leaves the other elements as they were, that is, zeros.

Now, if we say `Mz(3,5)=4`, we will set the third row and fifth column element.

```
>> Mz(3,5)=4   % 3rd row, 5th column

Mz =
```

```
        0        0        0        0        0
        0        0        0        1        0
        0        0        0        0        4
```

An alternative way to create a matrix would be to specify all elements of the matrix. The column elements should be separated by commas or spaces, and the row elements should be separated by semicolons. To recreate the above matrix, we can execute the following:

```
>> Mz=[ ...
0, 0, 0, 0, 0; ...
0, 0, 0, 1, 0; ...
0, 0, 0, 0, 4]

Mz =
        0        0        0        0        0
        0        0        0        1        0
        0        0        0        0        4
```

Notice the triple dot (...) mark; this tells MATLAB that the input will continue on the next line, and MATLAB should wait with the evaluation of the statement.

If a matrix has only one dimension, than it is often referred to as a *vector*. We can subdivide vectors to column vectors if they have $m \times 1$ dimensions, and row vectors, whose dimensions are $1 \times m$.

To create a row vector, we can, for example, type

```
>> % use comma to separate column elements
>> v=[1, 2, 3, 4, 5, 6, 7, 8]
v =
        1        2        3        4        5        6        7        8
>> % alternatively we can use spaces as separators
>> v=[1 2 3 4 5 6 7 8];
>> % or mix these two notations (NOT RECOMMENDED)
>> v=[1 2 3, 4, 5, 6 7 8]
v =
        1        2        3        4        5        6        7        8
```

A column vector constructed as

```
% use semicolon to separate row elements
>> vc=[1; 2; 3]
vc =
        1
        2
        3
```

There is yet one more way to create a matrix with prearranged column vectors. Here, we will use the already prepared column vector vc.

```
>> Mc=[vc, vc, vc]
Mc =

         1        1        1
         2        2        2
         3        3        3
```

In the following example, we will prepare the raw vector v and build the matrix Mv with simple arithmetic applied to the vector v.

```
v =
         1        2        3        4        5        6        7        8
>> Mv=[v; 2*v; 3*v]
Mv =
         1        2        3        4        5        6        7        8
         2        4        6        8       10       12       14       16
         3        6        9       12       15       18       21       24
```

2.5.2   Native matrix operations

MATLAB stands for **Mat**rix **Lab**oratory.[*] MATLAB was designed to process matrices efficiently and conveniently for the end user. We just saw an example where we performed multiplication on a vector to construct a matrix. Below, we show a couple more examples of operations with matrices.

Let's say we have the following matrix Mz.

```
Mz  =
         0        0        0        0        0
         0        0        0        1        0
         0        0        0        0        4
```

We can do the following operations without worrying how to perform the math on each element of our matrix.[†] Addition:

```
>> Mz+5
ans =
         5        5        5        5        5
         5        5        5        6        5
         5        5        5        5        9
```

---

[*] Deep inside of MATLAB, almost everything is a matrix, even a simple number. For example, let's assign the value 123 to the x variable: x=123. We can get the value of x by calling its name or by addressing it as the first element of a matrix. Operations 2*x, 2*x(1), 2*x(1,1) all produce the same result: 246.

[†] In many low-level programming languages (e.g., in C), this would be the programmer's job to implement correctly.

Multiplication:

```
>> Mz*2
ans =
     0     0     0     0     0
     0     0     0     2     0
     0     0     0     0     8
```

Note that for basic arithmetic with a matrix and a number, every element of the matrix gets the same treatment.

A function can take a matrix as the function argument and evaluate the value of the function for each matrix element.* We perform sin operation on matrix Mz:

```
>> sin(Mz)
ans =
     0        0        0        0        0
     0        0        0   0.8415        0
     0        0        0        0  -0.7568
```

We can add two matrices together:

```
>> Mz+Mz
ans =
     0     0     0     0     0
     0     0     0     2     0
     0     0     0     0     8
```

When two matrices participate in a mathematical operation, the rules are generally more complicated. For example, matrix multiplication is done according to the rules of linear algebra:

```
>> Mz*Mz'
ans =
     0     0     0
     0     1     0
     0     0    16
```

Here, Mz' corresponds to the complex conjugate transposed matrix Mz, that is, $Mz(i,j)' = Mz(j,i)$.* The complex conjugate does not do anything noticeable in this example, because all elements of Mz are real numbers.

2.5.2.1   Matrix element-wise arithmetic operators

There are special arithmetic operators that work on the elements of matrices, that is, they disregard linear algebra rules. Such *element-wise* operators start with the . (dot or period).

---

* Well, this is not always the case, but it is true for the basic mathematical functions. Some functions do some non-trivial transformations on the matrix element. For example, sum would add matrix elements column-wise and return a matrix with reduced dimensions.

Consider, for example, the element-wise multiplication operator .*

```
>> x=[1 2 3]
x = 1       2       3
>> y=[4   3   5]
y = 4       3       5
>> x.*y
ans =
            4       6      15
```

The result is obtained by multiplication of each element of x with the corresponding one from y. Note that the command x*y would produce an error, since the multiplication is not defined for the same type of vectors: x and y are both row vectors.

Yet one more example of the element-wise multiplication:

```
>> x=[1 2 3]
x = 1       2       3
>> x.*x % equivalent to x.^2 (see below)
     ans = 1      4       9
```

Here is an example of the element-wise division operator ./:

```
>> x./x
     ans = 1      1       1
```

Finally, an example of the element-wise power operator .^:

```
>> x.^2
     ans = 1      4       9
```

Let's move away from element-wise operations on vectors and perform such operations on two-dimensional matrices.

We define the matrix m to assist us.

```
>> m=[1,2,3; 4,5,6; 7,8,1]
m =
     1       2       3
     4       5       6
     7       8       1
```

In the following, we highlight differences of element-wise operators from the linear algebra equivalents.

Element-Wise Operator	Linear Algebra Rules Operator
Operator .∗	**Operator ∗**
| ```
>> m.*m
ans =
         1       4       9
        16      25      36
        49      64       1
``` | ```
>> m*m
ans =
        30      36      18
        66      81      48
        46      62      70
``` |
| **Operator .^** | **Operator ^ is not defined for two matrices** |
| ```
>> m.^m
ans =
         1              4             27
       256           3125          46656
    823543       16777216              1
``` | ```
% we expect this to fail
>> m^m
Error using    ^
Inputs must be a scalar
    and a square matrix.
``` |
| **Operator ./** | **Operator / works for two square matrices** |
| ```
% expect the matrix of
    ones
>> m./m
ans =
         1       1       1
         1       1       1
         1       1       1
``` | ```
% expect the unity matrix
>> m/m
ans =
         1       0       0
         0       1       0
         0       0       1
``` |

2.5.3 Strings as matrices

MATLAB stores a string as a one-dimensional array or matrix. You can access an individual character of a string by calling its number in the string.

```
>> s='hi there'
s =
  hi there

>> s(2)
ans =
  i

>> s(4)
ans =
  t
```

2.6 Colon (:) Operator

The : or *range* operator is **extremely useful**. It is commonly used in MATLAB to create vectors or matrix indexes. It usually takes the form `start:increment:stop` and creates a vector with the values [`start`, `start+1*increment`, . . . , `start+m*increment`], where m=1, 2, 3, 4, . . . and satisfies start≤start + m*increment≤stop, for the positive m.

It is much easier to understand what the : operator does by looking at the following examples.

```
>> v=5:2:12
v =
     5     7     9    11
```

The increment can be negative:

```
>> v2=12:-3:1
v2 =
    12     9     6     3
```

You can use the `start:stop` form with the default `increment = 1`:

```
>> v1=1:5
v1 =

     1     2     3     4     5
```

But there are some peculiarities. For example,

```
>> v3=5:1
v3 =
   Empty matrix: 1-by-0
```

produces a somewhat unexpected result. Naively, you would expect v3=5, but there are some built-in extra conditions. See them in the help browser, or execute `doc colon` in the command window.

2.6.1 Slicing matrices

We often need to select more than one element of a matrix, that is, a subset or a block of elements. Such an operation is called *slicing*; think about cutting a slice from a rectangular cake.

Here, we have a matrix Mv with size 3 × 8, and we want to choose all elements from columns 2, 5, and 6.

```
>> Mv
Mv =
     1     2     3     4     5     6     7     8
     2     4     6     8    10    12    14    16
     3     6     9    12    15    18    21    24
```

```
>> Mv(:,[2,5,6])
ans =
     2      5      6
     4     10     12
     6     15     18
```

The meaning of the : now is *choose all* relevant elements (rows in our case). Notice also that we used the vector ([2,5,6]) to specify the desired columns.

Similarly, we can take a substring of symbols from a longer string:

```
>> s='hi there'
s =
    hi there

>> s(4:8)
ans =
    there
```

2.7   Plotting

It is usually quite boring to stare at the bunch of numbers constituting matrices. It would be much more useful to show them in a graphical format.

Suppose we have a vector with values of x coordinates and want to plot $\sin(x)$ dependence. Below, we first create 10 linearly spaced x values in the range from 0 to 2π; this is done with the linspace function. Then, we find corresponding sin(x) values and assign them to the vector named y. Finally, we plot with the plot command.

```
>> x=linspace(0,2*pi,10)
x =
        0    0.6981    1.3963    2.0944    2.7925    3.4907
             4.1888  4.8869    5.5851    6.2832
>> y=sin(x)
y =
        0    0.6428    0.9848    0.8660    0.3420   -0.3420
    -0.8660   -0.9848   -0.6428   -0.0000
>> plot(x,y,'o') %  alternatively plot(x,sin(x),'o')
>> % every plot MUST have title, x and y labels
>> xlabel('x (radians)')
>> ylabel('sin(x)')
>> title('Plot of sin(x)')
```

The result is shown in Figure 2.6. Notice the third parameter in the plot command: 'o'. This is our way to specify the style of the points (circles in this

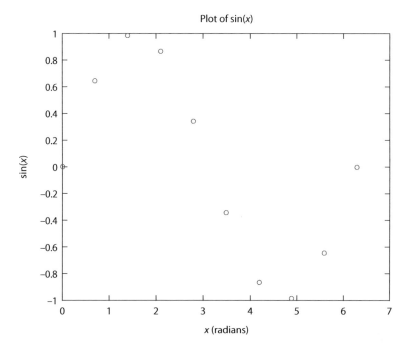

Figure 2.6 The plot of sin(x) with the default settings. Note that the font is quite small and difficult to read.

case). The plot command has many variations: you can specify point style, point color, and the style of the line connecting the points. Please read the relevant documentation for the full set of options. Also, notice that every professionally prepared plot should have a title and labeled axes. We did it with xlabel, ylabel, and title commands. We see the use for string type variables (which we discussed in Section 2.2.1): the strings are used for annotations.

Notice that the default font is rather small in Figure 2.6. Usually, this is fine on a personal display. However, such a small font size is not acceptable for publications. The command set(gca,'FontSize',24) sets the font to a larger value (24). It looks rather cryptic, but gca stands for Get the Current Axis object. This object is a collection of several plot properties. Out of these properties, we changed only the font size property, and the others are left intact. For the font setting command to take effect, we need to rerun the entire sequence of plot-related commands:

```
>> plot(x,y,'o')
>> set(gca,'FontSize',24);
>> xlabel('x (radians)')
>> ylabel('sin(x)')
>> title('Plot of sin(x)')
```

The result is depicted in Figure 2.7. The font is much larger now.

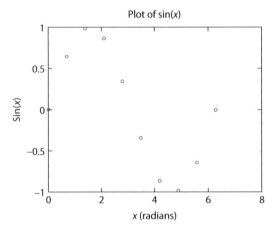

Figure 2.7 The plot of $\sin(x)$ with the increased font setting.

2.7.1 Saving plots to files

You can save your plots (and figures in general) either via the drop down menu associated with a figure or via execution of the print command.[*]

To save the figure in the pdf format, execute

```
>> print('-dpdf', 'sin_of_x')
```

This will generate the file 'sin_of_x.pdf'. Notice that MATLAB automatically added the appropriate file extension (pdf).

The −d switch designates the output format: pdf, ps, eps, png...

> **Words of wisdom**
>
> Do not save your plots in the JPEG format. It is a so-called lossy format, which saves a general appearance, and it is not designed to work with images that contain sharp transitions from one color to another. As a result, your labels will be blurry and hard to read, sharp lines will acquire strange artifacts, and so on.

Unfortunately, MATLAB generates pdf files with huge margins. Getting rid of these generally useless white spaces is not a trivial task.[†] Therefore, a pdf-formatted figure is not suitable for publications. The easiest short cut to obtain publication-quality images is probably to generate a png file, which has very tight bounds around the plot.

```
>> print('-dpng', '-r100', 'sin_of_x')
```

---

[*] As usual, the author urges the reader to read the documentation about the print command to learn the full set of its capabilities.

[†] The author spent quite a lot of time to find a way. If you search the web, you will see that the question "how to remove pdf margins in MATLAB?" is quite popular.

By default, the figure size is 8×6 inches;[*] the $-r$ switch sets the figure resolution in *dpi* (dots per inch). In this case, it is 100 dpi, so the resulting image will be 800×600 pixels. Feel free to increase or decrease the resolution parameter according to your needs.

2.8   Self-Study

Please refresh your memory of `plot`, `linspace`, and `print`, which were discussed earlier in Section 2.7. Additionally, consult relevant MATLAB help sections.

Problem 2.1
Plot the function $f(x) = \exp(-x^2/10) * \sin(x)$ for 400 linearly spaced points of x in the region from 0 to 2π. Points should be joined with solid lines.

Do not use any cycles or loops.

Problem 2.2
Plot functions x^2 and $x^3/2 + 0.5$ for 100 linearly spaced points of x in the region from -1 to $+1$. x^2 should be a red solid line and x^3 should be a black dashed line.

Do not use any cycles or loops.

---

[*] Yes, we are still using inches in the twenty-first century.

CHAPTER 3

Boolean Algebra, Conditional Statements, Loops

This chapter provides an overview on how to specify a conditional statement to a computer using MATLAB. It begins with an explanation of MATLAB Boolean logic with examples, use of comparison operators, and comparison with vectors. We proceed to show readers how to use "if-end-else" as well as how to use loops to prescribe repetitive tasks.

So far, the flow of calculations was direct. If we had a decision to make, we did it ourselves. This is fine for small calculation sessions, but we would like to delegate decision making to a computer. In other words, we need to learn how to specify a conditional statement to a computer. Mastering this concept is the goal for this chapter.

3.1 Boolean Algebra

Decision making requires the evaluation of a statement's truthness. In Boolean algebra, a statement (e.g., "the light is on") could be either *true* or *false*. This type of logic is traced back to Aristotelian times and is quite natural.* Consequently, a variable of Boolean type can have only two states or values:

- `false`: (MATLAB uses numerical 0 to indicate it.)

- `true`: (MATLAB uses numerical 1. Actually, everything but zero is treated as the true value as well.)

There are several logical operators that are used in Boolean algebra. Among them, the following three are the most fundamental.

- Logical *not*, MATLAB uses ~ (the tilde symbol) to indicate it

    `~true = false`

    `~false = true`

- Logical *and*, MATLAB uses &

$$A \ \& \ B = \begin{cases} \texttt{true, if A = true and B = true,} \\ \texttt{false, otherwise} \end{cases}$$

---

* There is also the so-called *fuzzy* logic, where we are sometimes in a "mixed" or undecided state. But this is beyond the scope of this book.

- Logical *or*, MATLAB uses | (the pipe symbol)

$$A \mid B = \begin{cases} \text{false, if } A = \text{false and } B = \text{false,} \\ \text{true, otherwise} \end{cases}$$

For any value of Z

```
~ Z & Z = false
```

Thus, the statement "Cats are animals and cats are not animals" is a false statement.

3.1.1 Boolean operators precedence in MATLAB

We list the logical operators according to MATLAB precedence: ~ has highest precedence, then &, and then |.

Consider the following example:

```
A | ~ B & C
```

We will add parentheses to indicate the precedence of operations (recall that expressions surrounded by parentheses are evaluated first). The above expression is equivalent to

```
A | ( (~B) & C)
```

Thus, for A = false, B = true, and C = true,

```
A | ~ B & C = false
```

3.1.2 MATLAB Boolean logic examples

Recall which numbers MATLAB treats as true and false to understand the following examples.

```
>> 123.3 & 12
ans = 1
```

```
>> ~ 1232e-6
ans = 0
```

Logical operations with matrices:

```
>> B=[1.22312, 0; 34.343, 12]
B =
    1.2231        0
   34.3430    12.0000
```

```
>> ~B
ans =
     0     1
     0     0
```

Let's address Hamlet's famous question: "To be or not to be?"

```
>> B|~B
ans =
     1     1
     1     1
```

We arrive at an answer which would certainly puzzle Hamlet: true. It is true for any values of B.

For two different matrices defined as follows,

```
>> B=[1.22312, 0; 34.343, 12]
B =
     1.2231         0
    34.3430    12.0000

>> A=[56, 655; 0, 24.4]
A =
    56.0000   655.0000
          0    24.4000
```

we get the following:

```
>> B&A
ans =
     1     0
     0     1
```

```
>> A|~B
ans =
     1     1
     0     1
```

3.2 Comparison Operators

MATLAB has the full set of numerical comparison operations shown in the following table.

Name	Math Notations	MATLAB's Notation	Comments
Equality	$=$	==	Note doubled equal signs
Non equality	$\neq$	~=	
Less	$<$	<	
Less or equal	$\leq$	<=	
Greater	$>$	>	
Greater or equal	$\geq$	>=	

The comparison operators allow us to perform *which element* and *choose* actions. This is demonstrated in Section 3.2.1.

3.2.1   Comparison with vectors

There are several examples with x defined as the following.

```
>> x=[1,2,3,4,5]
x =
     1    2    3    4    5
```

Let's have a look at the following statement: x >= 3. It is tempting to interpret it as "is x greater than or equal to 3?" But x has many elements. Some could be less than 3, and some could be greater. The question is ambiguous and not well defined, since we do not know which elements to use in comparison.

The correct way to interpret the statement x >= 3 is to read it as "which elements of x are greater than or equal to 3?"

```
>> x >= 3
ans =
     0    0    1    1    1
```

Note that the resulting vector is the same length as x, and the answer holds true or false in the corresponding position for each of the x elements.

Here is a more interesting use of the comparison operator: choose elements of x that are greater than or equal to 3.

```
>> x(x >= 3)
ans =
     3    4    5
```

The result is the subset of x.

3.2.2   Comparison with matrices

Now, let's define two matrices: A and B.

```
>> A=[1,2;3,4]
A =
1     2
3     4
```

```
>> B=[33,11;53,42]
B =
33        11
53        42
```

| Which elements of A are greater or equal to 2? | Choose elements of A that are greater or equal to 2. | Choose such elements of B where elements of A≥2. |

```
>> A>=2

ans =
       0      1
       1      1
```

```
>> A(A>=2)
ans =
       3
       2
       4
```

```
>> B(A>=2)
ans =
       53
       11
       42
```

Notice that the *choose* operation returns a column vector even if the input was a matrix.

3.3 Conditional Statements

3.3.1 The if-else-end statement

Finally, we are ready to program some conditional statements. An example of such a statement in a plain English is "if you are hungry, then eat some food; else, keep working." MATLAB's `if` expression is very similar to the human form:

```
if <hungry>
  eat some food
else
  keep working
end
```

Note that MATLAB does not use "then," since it is unnecessary. Even in English, we sometimes skip it: "if hungry, eat ...." Also, instead of a period, MATLAB uses the special keyword end.

A more formal definition of the if–else–end statement is

```
if <expression>
    this part is executed only when <expression> is true
else
    this part is executed only when <expression> is false
end
```

Note that if, else, and end are MATLAB's reserved keywords; thus, we cannot use them as variable names.

A fully MATLAB compatible example is

```
if (x>=0)
  y=sqrt(x);
else
  error('cannot do');
end
```

3.3.2 Short form of the "if" statement

There is often no need for the else clause. For example, "if you win a million dollars, then go party." There is a MATLAB equivalent for this statement:

```
if <expression>
    this part is executed only when <expression> is true
end
```

An example of this is

```
if (deviation<=0)
  exit;
end
```

3.4 Common Mistake with the Equality Statement

Have a look at the following code, and try to guess what value will be assigned to D after executing the if statement.

```
x=3; y=5;
if (x=y)
  D=4;
else
  D=2;
end
```

You might think that D is assigned to be 2, since x is not equal to y. But if you attempt to execute the above code, MATLAB will throw the error

```
if (x=y)
    |
Error: The expression to the left of the equals sign is
  not a valid target for an assignment.
```

This message looks quite cryptic, but it actually means that we attempted to use an assignment operator (=, i.e., single equal symbol) instead of the equality operator (==, i.e., double equal symbol).

3.5 Loops

3.5.1 The "while" loop

We often have to do a repetitive task: take a brick, put it into the wall, take a brick, put it into the wall, take a brick, put it into the wall, ... It would be silly to prescribe the action for every brick. So, we often create the assignment in a form of the loop: while the wall still needs to be finished, put bricks into the wall. MATLAB has the while–end loop to prescribe the repetitive work.

```
while <expression>
    this part (the loop body) is executed as long as the
        <expression> is true
end
```

The end at the bottom indicates the end of the prescription. It is not a signal to exit or finish the loop.

As an example, let's add the integer numbers from 1 to 10.

```
s=0; i=1;
while (i<=10)
  s=s+i;
  i=i+1;
end
```

Now s holds the value of the sum.

```
>> s
   s = 55
```

It is 55, as expected.

The while loop is extremely useful, but it is not guaranteed to finish. If the conditional statement in the loop body is a bit more complicated, it may be impossible to predict whether the loop will finish or not.

Consequently, it is easy to forget to create the proper exit condition. Have a look at the following code:

```
s=0; i=1;
while (i<=10)
  s=s+i;
end
```

At first glance, it looks exactly like the previous code calculating the sum of integers from 1 to 10. But if you attempt to run it, the computer will crunch numbers as long as it is on. If you executed the above code or need to *stop the execution* of your program, just press two keys together: Ctrl and c. Now, let's see what the problem is. There was a forgotten statement that updates i, so it was always 1. Thus, the i<=10 condition was always true, and the loop was doomed to run forever.

Words of wisdom

When you work with a while loop, program the statements that are in charge of the loop exit **first**.

3.5.2 Special commands "break" and "continue"

There are situations when we would like to stop the loop in the middle of its body or when some of the conditions are fulfilled. For this purpose, there is a special command, break, which stops the execution of the loop. Let's see how can we count numbers from 1 to 10 another way using the break command.

```
s=0; i=1;
while (i > 0)
  s=s+i;
  i=i+1;
  if (i > 10)
    break;
  end
end
```

Yet another special command is continue. It interrupts the body of the loop execution and starts it from the beginning. Let's demonstrate it with the same problem of counting from 1 to 10.

```
s=0; i=1;
while (i > 0)
  s=s+i;
  i=i+1;
  if (i < 11)
    continue;
  end
  break;
end
```

These examples look more and more complicated, but there are situations where the use of continue or break will actually simplify the code.

3.5.3 The "for" loop

The while loop is sufficient for any programming, but, as mentioned in the previous subsection, it requires careful tracking and programming the exit condition. The for loop does not have this issue, though it is less general.

```
for <variable>=<expression>
    the body of the loop will be executed with <variable>
    set consequently to each column of the <expression>
end
```

A very common idiom involving for looks like for i=initial:final. In this case, we can read it as: for each integer i spanning from initial to final, do something. We will demonstrate it on the same counting example.

```
s=0;
for i=1:10
     s=s+i;
end
```

The s holds 55 again.

The numbers in the for loop assignments do not need to be consequent. Here is one more example, which counts the sum of all elements of x.

```
sum=0;
x=[1,3,5,6]
for v=x
   sum=sum+v;
end
```

```
>> sum
sum =
    15
```

The for loops are guaranteed to complete after a predictable number of iterations (the number of columns in the *<expression>*). Nevertheless, the afore-mentioned commands break and continue work in for loops as well, and we are free to use them to interrupt or redirect the loop flow.

3.5.3.1 Series implementation example
Let's implement the following series using MATLAB:

$$S = \sum_{k=1}^{k \leq 100} a_k \tag{3.1}$$

where:

$$a_k = k^{-k}$$
$$a_k \geq 10^{-5}$$

Once we start to program a relatively complex expression, there are multiple ways to reach the same results. However, you probably should aim to find the most concise code.

In the following, we demonstrate several implementations of Equation 3.1.

```
S=0; k=1;
while( (k<=100) & (k^-k
        >= 1e-5) )
  S=S+k^-k;
  k=k+1;
end
%
%
%
```

```
S=0; k=1;
while( k<=100 )
  a_k=k^-k;
  if (a_k < 1e-5)
     break;
  end
  S=S+a_k;
  k=k+1;
end
```

```
S=0;
for k=1:100
  a_k=k^-k;
  if (a_k < 1e-5)
     break;
  end
  S=S+a_k;
end
%
```

```
>> S
S =
   1.2913
```

```
>> S
S =
   1.2913
```

```
>> S
S =
   1.2913
```

As you can see, all three ways give the same result, but the implementation on the left seems to be more clear.

If you have programmed before, loops are a very natural way to do things. However, you should aim to use the matrices operation capabilities of MATLAB whenever it is possible.

Let's see how we can do Equation 3.1 without a loop at all.

```
>> k=1:100;
>> a_k=k.^-k; % a_k is the vector
>> S=sum(a_k(a_k>=1e-5))
S =
   1.2913
```

In this code, we used the "choose elements" construct and the built-in sum function.

Words of wisdom

If you care about speed, avoid loops as much as possible and use Matlab's capabilities to operate on matrices. This frequently produces a more concise code as well.

3.6   Self-Study

Problem 3.1
Do the problems at the end of Chapter 4 listed in Section 4.6. If you are not sure what function is, just make a script, that is, a sequence of commands.

Functions, Scripts, and Good Programming Practice

The art of programming is the ability to translate from human notation to one that a computer understands. In the spirit of a gradual increase of complexity, we always start with mathematical notation, which serves as the bridge between human language and computer (programming) language. Essentially, mathematical notation is the universal language from which we can always go to an arbitrary programming language, including MATLAB. This chapter addresses functions, scripts, and basic good programming practices. It begins with some motivational examples and shows how to run test cases to check your solutions, making sure they are realistic. We conclude the chapter by discussing recursive and anonymous functions.

4.1 Motivational Examples

Before we jump to functions and scripts, we will cover two examples: the first from personal finances and the second from physics.

4.1.1 Bank interest rate problem

Suppose someone desires to buy a car with price Mc two years from now, and this person currently has a starting amount of money Ms. What interest rate is required to grow the starting sum to the required one?

As usual, our first job is to translate the problem to the language of mathematics and then convert it to a programming problem. Usually, the interest rate is given as a percentage (p) by which the account grows every year. Well, percentages are really for accountants, and everyone else uses fractions of 100, that is, $r = p/100$, so we say that our initial investment grows by $1 + r$ every year. Thus, in two years it will grow as $(1 + r)^2$, and we finally can form the following equation:

$$Ms \times (1 + r)^2 = Mc$$

Let's go back to using fractions.

$$Ms \times (1 + p/100)^2 = Mc$$

Now, we expand the equation:

$$1 + 2\frac{p}{100} + \frac{p^2}{100^2} = \frac{Mc}{Ms}$$

With the following relabeling $p \to x$, $1/100^2 \to a$, $1/50 \to b$, and $(1 - Mc/Ms) \to c$, the canonical quadratic equation is easily seen:

$$ax^2 + bx + c = 0 \tag{4.1}$$

For now, we will postpone the solution of this equation and see one more problem where we need to solve such an equation as well.

4.1.2   Time of flight problem

A fireworks pyrotechnician would like to ignite a charge at the height (h) to maximize the visibility of the flare and to synchronize its position with other flares. The firework's shell leaves a gun with vertical velocity (v). We need to determine the delay time to which we must set the ignition timer, that is, find how long it takes for the shell to reach the desired height (h) with respect to the firing position (see Figure 4.1). Again, first we need to translate the problem from the language of physics to mathematics. Assuming that the gun is not too powerful, we can treat the acceleration due to gravity (g) as a constant at all shell heights. We will also neglect air resistance. The shell's vertical position (y) versus flight time (t) is governed by the equation of motion with constant acceleration and expressed as $y(t) = y_0 + vt - gt^2/2$, where y_0 is the height of the shell at the firing time, that is, $y(0) = y_0$. So, we need to solve the following equation:

$$h = y(t) - y_0$$

Substituting known parameters, we obtain

$$h = vt - gt^2/2 \tag{4.2}$$

Finally, we convert this equation to the canonical quadratic Equation 4.1 with the following substitutions: $t \to x$, $-g/2 \to a$, $v \to b$, and $-h \to c$.

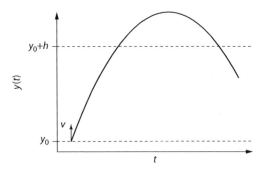

Figure 4.1   The firework rocket's height dependence on time.

4.2 Scripts

So far, when we interacted with MATLAB, we typed our commands in the command window in sequence. However, if we need to execute repetitive sequences of commands, it is much more convenient and, more importantly, less error prone to create a separate file with such commands, that is, the script file.

The name of the script file is arbitrary,[*] but it should end with .m to mark this file as executable by MATLAB. The script file by itself is nothing but a simple text file, which can be modified not only by MATLAB's built-in editor but by any text editor as well.

Let's say we have a script file labeled 'script1.m'. To execute its content, we need to type its name without .m in the command prompt, that is, script1. MATLAB will go over all commands listed in the script and execute it as if we were typing them one by one. An important consequence of this is that it will modify our workspace variables, as we will soon see.

4.2.1 Quadratic equation solver script

After the formal definitions of a script, mentioned previously, we address our problem of solving the quadratic equation

$$ax^2 + bx + c = 0$$

with the script (i.e., the program).

Before we start programming, we still need to spend time in the realm of mathematics. The quadratic equation has in general two roots, x_1 and x_2, which are given by the following equation:

$$x_{1,2} = \frac{-b \pm \sqrt{b^2 - 4ac}}{2a}$$

The MATLAB-compatible representation of this equation is shown in the following listing, which we will save into the file 'quadSolvScrpt.m'.

Listing 4.1 quadSolvScrpt.m (available at http://physics.wm.edu/
programming_with_MATLAB_book/./ch_functions_and_scripts/code/
quadSolvScrpt.m)

```
% solve quadratic equation   a*x^2 + b*x + c = 0
x1 = ( - b - sqrt( b^2 - 4*a*c ) ) / (2*a)
x2 = ( - b + sqrt( b^2 - 4*a*c ) ) / (2*a)
```

---

[*] Though a script filename is arbitrary, it is a good idea to choose a filename that reflects the purpose of the script.

Words of wisdom

Put many comments in your code, even if you are the only intended reader. Trust the author; about two weeks after completion of your script, you likely will not be able to recall the exact purpose of the code or why a particular programming decision was made.

Now, let's define coefficients *a*, *b*, and *c* and run our script to find the roots of the quadratic equation. We need to execute the following sequence:

```
>> a = 2; b =-8; c = 6;
>> quadSolvScrpt
x1 =
     1
x2 =
     3
```

Notice that MATLAB creates variables x1 and x2 in the workspace and also displays their values as the output of the script. As usual, if we want to suppress the result of any MATLAB statement execution, we need to screen the statement by terminating it with ;.

If we would like to find the solution of the quadratic equation with different coefficients, we just need to redefine them and run our script again.

```
>> a = 1; b =3; c = 2;
>> quadSolvScrpt
x1 =
    -2
x2 =
    -1
```

Note that our old values of x1 and x2 are overwritten with new ones.

The ability of scripts to overwrite the workspace variables can be quite handy. For example, one might want to create a set of coefficients or universal constants for further use during a MATLAB session. See the following sample script, which sets some universal constants.

Listing 4.2 `universal_constants.m` (available at `http://physics.wm.edu/programming_with_MATLAB_book/./ch_functions_and_scripts/code/universal_constants.m`)

```
% the following are in SI units
g = 9.8;  % acceleration due to gravity
c = 299792458; % speed of light
G = 6.67384e-11; % gravitational constant
```

We can use this script to easily find, for example, what height the firework's shell (which we discussed in Section 4.1.2) will reach at a given time $t = 10\,\text{s}$, assuming that the initial vertical velocity is $v = 142$ m/s. Since g is defined in the script, to calculate the height according to Equation 4.2, all we need to do is to execute the following:

```
>> universal_constants
>> t=10; v= 142;
>> h = v*t - (g*t^2)/2
h =
   930
```

4.3   Functions

In general, it is considered bad practice to use scripts, as their ability to modify workspace variables is actually a downside when working on a complex program. It would be nice to use code such that after execution, only the results are provided without the workspace being affected.

In MATLAB, this is done via *function*. A function is a file that contains the structure

```
function [out1, out2, ..., outN] = function_name (arg1, arg2, ..., argN)
        % optional but strongly recommended function description
        set of expressions of the function body
end
```

Note that the file name must end with '.m', and the leading part of it must be the same as the function name, that is, 'function_name.m'.

Words of wisdom

White space, consisting of tabulations and spaces in front of any line, is purely cosmetic but greatly improves the readability of a program.

A function might accept several input arguments or parameters. Their names can be arbitrary, but it is a good idea to give them more meaningful names, that is, not just arg1, arg2, ..., argN. For example, our quadratic solver could have much better names for input parameters than a, b, and c. Similarly, a function can return several parameters, but they must be listed in the square brackets [] after the function keyword. Their order is completely arbitrary, like assignment names. Again, it is much better to use something like x1 and x2 instead of out1 and out2. The only requirement is that the return parameters need to be assigned somewhere in the body of the function, but do not worry—MATLAB will prompt you if you forget about this.

Words of wisdom

By the way, many built-in MATLAB functions are implemented as '.m' files following these conventions. You can learn good techniques and tricks of MATLAB by reading these files.

4.3.1 Quadratic equation solver function

Usually, when I develop a function, I start with a script, debug it, and then wrap it with the function related statements. We will modify our quadratic equation solver script to become a valid MATLAB function.

Listing 4.3 quadSolverSimple.m (available at http://physics.wm.edu/ programming_with_MATLAB_book/./ch_functions_and_scripts/code/ quadSolverSimple.m)

```
function [x1, x2] = quadSolverSimple( a, b, c )
% solve quadratic equation  a*x^2 + b*x + c = 0
x1 = ( - b - sqrt( b^2 - 4*a*c ) ) / (2*a);
x2 = ( - b + sqrt( b^2 - 4*a*c ) ) / (2*a);
end
```

Note that we need to save it to 'quadSolverSimple.m'. Now, let's see how to use our function.

```
>> a = 1; b =3; c = 2;
>> [x1, x2] = quadSolverSimple(a,b,c)
x1 =
     -2
x2 =
     -1
```

Now, we highlight some very important features of MATLAB functions.

```
>> x1=56;
>> clear x2
>> a = 1; b =3; cc = 2; c = 200;
>> [xx1, xx2] = quadSolverSimple(a,b,cc)
xx1 =
     -2
xx2 =
     -1
>> x1
x1 =
     56
>> x2
Undefined function or variable 'x2'.
```

Note that at the very first line, we assign x1 to be 56, and then at the very end, we check the value of the x1 variable, which is still 56. As our function listing shows, MATLAB assigns the x1 variable as one of the roots of the quadratic equation inside of the function body. However, this assignment does not affect the x1 variable in the workspace. The second thing to note is that the function assigned xx1 and xx2 variables in our workspace; this is because we asked to assign these variables. This is another feature of MATLAB functions—they assign variables in the workspace in the same order and to variables that the user asked for during the function call. Notice also that since we clear x2 before the function call, there is still no assigned x2 variable after the execution of the function, though the quadSolverSimple function uses it internally for its own needs. Finally, notice that the function clearly did not use the assigned value of c = 200. Instead, it used the value of cc = 2, which we provided as the third argument to the executed function. So, we note that the names of parameters during the function call are irrelevant, and only their position is important.

We can simply remember this: *what happens in the function stays in the function*, and what comes out of the function goes into the return variable placeholders.

4.4 Good Programming Practice

If this is your first time studying programming, then you can skip this section. Come back to it later, once you are fluent with the basics of functions and scripts.

Here, we discuss how to write programs and functions in a robust and manageable way based on our simple quadratic equation solver function.

4.4.1 Simplify the code

Let's have a look at the original quadSolverSimple function Listing 4.3. At first glance, everything looks just fine, but there is a part of the code that repeats twice, that is, calculation of b^2 − 4*a*c. MATLAB wastes cycles to calculate it a second time, and, more importantly, the expression is the definition of the discriminant of our equation, which can be very useful (we will see it very soon). Additionally, if we find a typo in our code, it is very likely that we will forget to fix it at the second occurrence of such code, especially if it is separated by more than a few lines. So, we transform our function to the following:

Listing 4.4 quadSolverImproved.m (available at http://physics.wm.edu/programming_with_MATLAB_book/./ch_functions_and_scripts/code/quadSolverImproved.m)

```
function [x1, x2] = quadSolverImproved( a, b, c )
% solve quadratic equation  a*x^2 + b*x + c = 0
D = b^2 - 4*a*c;
x1 = ( - b - sqrt( D ) ) / (2*a);
x2 = ( - b + sqrt( D ) ) / (2*a);
end
```

4.4.2 Try to foresee unexpected behavior

This looks much better, but what if our discriminant is negative? Then, we cannot extract the square root, and the function will fail (technically, we can do it, but this involves manipulation with complex numbers, and we pretend that this is illegal). Therefore, we need to check whether the discriminant is positive and produce a meaningful error message otherwise. For the latter, we will use MATLAB's error function, which stops program execution and produces a user-defined message in the command window.

Listing 4.5 `quadSolverImproved2nd.m` (available at `http://physics.wm.edu/` `programming_with_MATLAB_book/./ch_functions_and_scripts/code/` `quadSolverImproved2nd.m`)

```matlab
function [x1, x2] = quadSolverImproved2nd( a, b, c )
% solve quadratic equation  a*x^2 + b*x + c = 0
D = b^2 - 4*a*c;
if ( D < 0 )
    error('Discriminant is negative: cannot find real
        roots');
end
x1 = ( - b - sqrt( D ) ) / (2*a);
x2 = ( - b + sqrt( D ) ) / (2*a);
end
```

4.4.3 Run test cases

At this point, our quadratic solver looks quite polished, and nothing can possibly go wrong with it. Right? As soon as you come to this conclusion, a loud bell should ring inside of your mind: I am getting too comfortable; thus, it is time to check. **Never release or use a code which you did not check.**[*]

---

[*] This is true both for your programs and especially for code that you have received from others, even if this source is very trustworthy and reputable. You might think that big software companies have tons of experience and produce "bulletproof" quality code. Their experience does not guarantee error-free code. **Absolutely all** software packages that this author has seen in his life come with a clause similar to this: "THERE IS NO WARRANTY FOR THE PROGRAM. THE ENTIRE RISK AS TO THE QUALITY AND PERFORMANCE OF THE PROGRAM IS WITH YOU. SHOULD THE PROGRAM PROVE DEFECTIVE, YOU ASSUME THE COST OF ALL NECESSARY SERVICING, REPAIR OR CORRECTION" (excerpt from GPL license). It comes in capital letters, so we should take it seriously. Also, recall the MATLAB license agreement, which you agreed to at the beginning of MATLAB usage. MATLAB expresses it in a slightly less straightforward way, but the meaning is the same: "any and all Programs, Documentation, and Software Maintenance Services are delivered 'as is' and MathWorks makes and the Licensee receives no additional express or implied warranties."

   This seems like a very long paragraph with more emphasized phrases than in the rest of this book; it also seems to be irrelevant to the art of programming. But this author has spent a lot of his time pulling hair from his head trying to see mistakes in his code and finding the problem rooted in someone else's code (I do not imply that I produce error-free code). So, **trust but verify** everyone.

Equipped with this motto, we will start with testing our own code; this is commonly called *running test cases*. Ideally, testing should cover all possible input parameters, although this is clearly impossible. We should check that our program produces the correct answer in at least one case, and we need to verify this answer in a somewhat independent way. Yes, often it means the use of paper and pencil. We also need to poke our program with somewhat random input parameters to see how robust it is.

First, we double check the correctness of the function. We will use simple enough numbers that we can do it in our head. Actually, we already did it in previous test runs during this chapter, but you can never have enough tests. So, we will do it one more time.

```
>> a = 1; b = -3; c = -4;
>> [x1,x2] = quadSolverImproved2nd(a, b, c)
x1 =
     -1
x2 =
      4
```

It is easy to see that $(x - 4) * (x + 1) = x^2 - 3x - 4 = 0$, that is, the produced roots, indeed, satisfy the equation with the same $a = 1$, $b = -3$, and $c = -4$ coefficients. By the way, **do not use the same code for correctness verification; use some independent method**.

Now, we check the case when the discriminant is negative:

```
>> a = 1; b = 3; c = 4;
>> [x1,x2] = quadSolverImproved2nd(a, b, c)
Error using quadSolverImproved2nd (line 5)
Discriminant is negative: cannot find real roots
```

Excellent! As expected, the program terminates with the proper error message.

4.4.4   Check and sanitize input arguments

One more test:

```
>> a = 0; b = 4; c = 4;
>> [x1,x2] = quadSolverImproved2nd(a, b, c)
x1 =
   -Inf
x2 =
    NaN
```

Wow! There is no way that an equation with $a = 0$ (simplified to $bx + c = 4x + 4 = 0$) has one root equal to infinity and the other root being NaN, which stands for "not a number." We do not need a calculator to see that the root of this equation is -1. What is going on?

Let's closely examine our code in Listing 4.5; the problem part is division by $2a$, which is actually 0 in this case. This operation is undefined, but, unfortunately, MATLAB is trying to be smart and not produce an error message. Sometimes, this is welcome behavior, but now it is not. So, we need to be in control: intercept the case of $a = 0$ and handle it separately, producing solutions $x1 = x2 = -c/b$. It is easy to see that we need to handle the case $a = b = 0$ as well. So, our final quadratic solver function will be

Listing 4.6 quadSolverFinal.m (available at http://physics.wm.edu/ programming_with_MATLAB_book/./ch_functions_and_scripts/code/ quadSolverFinal.m)

```
function [x1, x2] = quadSolverFinal( a, b, c )
% solve quadratic equation  a*x^2 + b*x + c = 0

% ALWAYS check and sanitize input parameters
if ( (a == 0) & (b == 0) )
    error('a==0 and b==0: impossible to find roots');
end

if ( (a == 0) & (b ~= 0) )
    % special case: we essentially solve b*x = -c
    x1 = -c/b;
    x2=x1;
else
    D = b^2 - 4*a*c; % Discriminant of the equation
    if ( D < 0 )
        error('Discriminant is negative: no real roots');
    end
    x1 = ( - b - sqrt( D ) ) / (2*a);
    x2 = ( - b + sqrt( D ) ) / (2*a);
end
end
```

4.4.5   Is the solution realistic?

This was a lot of work to make a quite simple function perform to specifications. Now, let's use our code to solve our motivational examples.

We start with the interest rate problem described in Section 4.1.1. Suppose that we initially had $10,000$ (Ms = 10000), and the desired final account value is $20,000$ (Mc = 20000), thus:

```
>> a = 1/100^2; b = 1/50; c = 1 - 20000/10000;
>> [p1,p2] = quadSolverFinal(a, b, c)
p1 =
 -241.4214
```

```
p2 =
   41.4214
```

At first glance, everything is fine, since we obtain two solutions for required interest rate -241.4% and 41.4%. But if we look more closely, we realize that negative percentage means that we owe to the bank after each year, so the account value will decrease every year—which is opposite to our desire to grow money.

What is the reason for such an "unphysical" solution? The real meaning of the problem was lost in translation to mathematical form. Once we have the $(1 + r)^2$ term, the computer does not care whether the squared number is negative or positive, since it produces the valid equation root. But we humans do care!

We have just learned one more important lesson: **it is up to the human to decide whether the solution is valid**. Never blindly trust a solution produced by a computer! They do not care about the reality or the "physics" of the problem.

4.4.6 Summary of good programming practice

- Foresee problem spots
- Sanitize and check input arguments
- Put a lot of comments in your code
- **Run test cases**
- Check the meaning of the solutions and exclude unrealistic ones
- Fix problem spots and repeat from the top

4.5 Recursive and Anonymous Functions

Before we move on, we need to consider a couple of special function use cases.

4.5.1 Recursive functions

Functions can call other functions (this is not a big surprise; otherwise, they would be quite useless), and they can call themselves, which is called *recursion*. If we go into detail, there is a limit to how many times a function can call itself. This is due to the limited memory size of a computer, since every function call requires the computer to reserve some amount of memory space to recall it later.

Let's revisit the account growth problem that we discussed in Section 4.1.1. Now, we would like to find the account value including interest after a certain number of years. The account value (Av) after N years is equal to the account value in the previous year ($N - 1$) multiplied by the growth coefficient $(1 + p/100)$. Assuming that we initially invested an amount of money equal to Ms, we can calculate the final account value according to

$$Av(N) = \begin{cases} Ms & \text{if } N = 0 \\ (1 + p/100) \times Av(N-1) & \text{if } N > 0 \end{cases} \qquad (4.3)$$

This equation resembles a typical recursive function, where the function calls itself to calculate the final value. The MATLAB implementation of such a function is as follows:

Listing 4.7   accountValue.m (available at http://physics.wm.edu/
programming_with_MATLAB_book/./ch_functions_and_scripts/code/
accountValue.m)

```
function Av = accountValue( Ms, p, N)
% calculates grows of the initial account value (Ms)
% in the given amount of years (N)
% for the bank interest percentage (p)

% We sanitize input to ensure that stop condition is possible
    if ( N < 0 )
        error('Provide positive and integer N value');
    end
    if ( N ~= floor ( N ) )
        error ('N is not an integer number');
    end

% Do we meet stop condition?
    if ( N == 0 )
        Av = Ms;
        return;
    end

    Av = (1+p/100)*accountValue( Ms, p, N-1 );
end
```

Let's see how the initial sum $Ms = \$535$ grows in 10 years if the account growth percentage is 5.

```
>> Ms=535; p=5; N=10; accountValue(Ms, p, N)
ans =
  871.4586
```

4.5.2 Anonymous functions

Anonymous functions look very confusing at first, but they are useful in cases when one function should call another, or if you need to have a short-term-use function, which is simple enough to fit in one line.

It is easier to start with an example. Suppose for some calculations you need the following function: $g(x) = x^2 + 45$. It is clearly very simple and would probably be used during only one session, and thus, there is no point in creating a full-blown .m file for this function. So, we define an anonymous function.

```
>> g = @(x) x^2 + 45;

>> g(0)
ans =
      45
>> g(2)
ans =
      49
```

The definition of the anonymous function g happens in the first line of this listing; the rest is just a few examples to prove that it is working properly. The @ symbol indicates that we are defining a *handle*** to the function of one variable x, as indicated by @(x), and the rest is simply the function body, which must consist of just one MATLAB statement resulting in a *single* output.

An anonymous function can be a function of many variables, as shown here:

```
>> h = @(x,y) x^2 - y + cos(y);
>> h(1, pi/2)
ans =
    -0.5708
```

Anonymous functions are probably the most useful when you want to define a function that uses some other function with some of the input parameters as a constant. For example, we can "slice" $h(x, y)$ along the x dimension for fixed $y = 0$, that is, define $h1(x) = h(x, 0)$:

```
>> h1 = @(x) h(x,0);
>> h1(1)
ans =
       2
```

Another useful property of anonymous functions is their ability to use variables defined in the workspace at the time of definition.

```
>> offset = 10; s = @(x) x + offset;
>> clear offset
>> s(1)
ans =
      11
```

---

* Handle is a special variable type. It gives MATLAB a way to store and address a function.

Note that in this transcript, the `offset` variable was cleared at the time of the s function's execution, yet it still works, since MATLAB already used the value of the variable when we defined the function.

We can also evaluate

$$\int_0^{10} s(x)dx$$

with the help of MATLAB's built-in function `integral`:

```
>> integral(s,0,10)
ans =
   150
```

Words of wisdom

Avoid using scripts. Instead, convert them into functions. This is much safer in the long run, since you can execute a function without worrying that it may affect or change something in your workspace in an unpredictable way.

4.6   Self-Study

Problem 4.1

Write a script that calculates

$$1 + \sum_{i=1}^{N} \frac{1}{x^i}$$

for $N = 10$ and $x = 0.1$.

Use loops as much as you wish from now on.

Problem 4.2

Write a script that calculates for $N = 100$

$$S_N = \sum_{k=1}^{N} a_k$$

where:

     $a_k = 1/k^{2k}$ for odd k
     $a_k = 1/k^{3k}$ for even k

     Hint: you may find the mod function useful to check for even and odd numbers.

Problem 4.3

Write a function `mycos` that calculates the value of a $\cos(x)$ at the given point x via the Taylor series up to N members. Define your function as
```
function cosValue = mycos(x, N)
```
Does it handle well the situation with large x values? Take $x = 10\pi$, for example. How far do you need to expand the Taylor series to get absolute precision of 10^{-4}, what value of N do you find reasonable (no need to state it beyond one significant digit), and why is this?

Problem 4.4

Download the data file `'hwOhmLaw.dat'`.* It represents the result of someone's attempt to find the resistance of a sample via measuring voltage drop (V), which is the data in the first column, and current (I), which is the data in the second column, passing through the resistor in the same conditions. Judging by the number of samples, it was an automated measurement.

- Using Ohm's law $R = V/I$, find the resistance (R) of this sample (no need to print it out for each point at this step).

- Estimate the resistance of the sample (i.e., find the average resistance) and estimate the error bars of this estimate (i.e., find the standard deviation).

For standard deviation, use the following definition:

$$\sigma(x) = \sqrt{\frac{1}{N-1}\sum_{i=1}^{N}(x_i - \bar{x})^2}$$

where:
  x is the set (vector) of data points
  $\bar{x}$ its average (mean)
  N is the number of the points in the set.

Do not use standard built-in `mean` and `std` functions in your solution. You need to make your own code to do it. But feel free to test against these MATLAB functions.

Note: for help, read MATLAB's `std`; you might want to know about it.

Problem 4.5

Imagine you are working with an old computer that does not have the built-in multiplication operation. Program the `mult(x,y)` function, which returns the equivalent of `x*y` for two integer numbers `x` and `y` (either one can be negative, positive, or zero). Do not use the `*` operator of MATLAB. You can use loops, conditions, `+`, or `−` operators. Define your function as
```
function product=mult(x,y)
```

---

* The file is available at `http://physics.wm.edu/programming_with_MATLAB_book/./`
  `ch_fitting/data/hwOhmLaw.dat`

Part II

Solving Everyday Problems with MATLAB

CHAPTER 5

Solving Systems of Linear Algebraic Equations

This chapter begins the second section addressing the use of MATLAB for solving everyday problems. It begins by presenting an example of a child's mobile and then uses built-in MATLAB solvers to explore various available methods (e.g., inverse matrix method and those that do not require inverse matrix calculation) for finding the solution. We then present another example, of the Wheatstone bridge circuit.

5.1 The Mobile Problem

We set up a problem by considering the following situation. Someone has provided us with six figures and three rods to make a child's mobile. We need to calculate the positions of suspension points (i.e., the lengths $x_1, x_2, \ldots, x_6$) to have a balanced system. Our mobile will look like the one shown in Figure 5.1. A good-looking mobile should be in a state of equilibrium, that is, all of the suspending arms must be close to horizontal. Here, we put our physicist hat on and will treat the suspended figures only as simple weights $w_1, w_2, \ldots, w_6$. We do not care if they are fish, clouds, or anything artistic. In the same spirit, we assume that rods with known lengths L_{12}, L_{34}, and L_{56} are weightless.

If the system is in equilibrium, torque must be zero at every pivot point. This forms the following equations:

$$w_1 x_1 - (w_2 + w_3 + w_4 + w_5 + w_6)x_2 = 0 \tag{5.1}$$

$$w_3 x_3 - (w_4 + w_5 + w_6)x_4 = 0 \tag{5.2}$$

$$w_5 x_5 - w_6 x_6 = 0 \tag{5.3}$$

We need three more equations. We note that $x_1, x_2, \ldots, x_6$ should add up to the corresponding rod length:

$$x_1 + x_2 = L_{12} \tag{5.4}$$

$$x_3 + x_4 = L_{34} \tag{5.5}$$

$$x_5 + x_6 = L_{56} \tag{5.6}$$

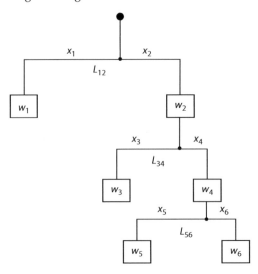

Figure 5.1   A mobile sketch.

Let's define $w_{26} = w_2 + w_3 + w_4 + w_5 + w_6$ and $w_{46} = w_4 + w_5 + w_6$ to simplify Equation 5.1 through Equation 5.3 and write the full system of equations

$$w_1 x_1 - w_{26} x_2 = 0 \tag{5.7}$$

$$w_3 x_3 - w_{46} x_4 = 0 \tag{5.8}$$

$$w_5 x_5 - w_6 x_6 = 0 \tag{5.9}$$

$$x_1 + x_2 = L_{12} \tag{5.10}$$

$$x_3 + x_4 = L_{34} \tag{5.11}$$

$$x_5 + x_6 = L_{56} \tag{5.12}$$

Now, let's spell out every equation, so it contains all $x_1, x_2, \ldots, x_6$, even if there are zero coefficients in front of some of them

$$
\begin{aligned}
w_1 x_1 - w_{26} x_2 + 0 x_3 + 0 x_4 + 0 x_5 + 0 x_6 &= 0 \\
0 x_1 + 0 x_2 + w_3 x_3 - w_{46} x_4 + 0 x_5 + 0 x_6 &= 0 \\
0 x_1 + 0 x_2 + 0 x_3 + 0 x_4 + w_5 x_5 - w_6 x_6 &= 0 \\
1 x_1 + 1 x_2 + 0 x_3 + 0 x_4 + 0 x_5 + 0 x_6 &= L_{12} \\
0 x_1 + 0 x_2 + 1 x_3 + 1 x_4 + 0 x_5 + 0 x_6 &= L_{34} \\
0 x_1 + 0 x_2 + 0 x_3 + 0 x_4 + 1 x_5 + 1 x_6 &= L_{56}
\end{aligned}
\tag{5.13}
$$

Now, we can see the structure and rewrite the system of equations in the canonical form

---

Matrix form of the system of linear equations

$$\mathbf{A}\mathbf{x} = \mathbf{B} \tag{5.14}$$

which is the shorthand notation for the following

$$\sum_j A_{ij} x_j = B_i \tag{5.15}$$

We spell out **A**, **x**, and **B**:

$$\begin{pmatrix} w_1 & -w_{26} & 0 & 0 & 0 & 0 \\ 0 & 0 & w_3 & -w_{46} & 0 & 0 \\ 0 & 0 & 0 & 0 & w_5 & -w_6 \\ 1 & 1 & 0 & 0 & 0 & 0 \\ 0 & 0 & 1 & 1 & 0 & 0 \\ 0 & 0 & 0 & 0 & 1 & 1 \end{pmatrix} \begin{pmatrix} x_1 \\ x_2 \\ x_3 \\ x_4 \\ x_5 \\ x_6 \end{pmatrix} = \begin{pmatrix} 0 \\ 0 \\ 0 \\ L_{12} \\ L_{34} \\ L_{56} \end{pmatrix} \tag{5.16}$$

We delay further discussion of the mobile problem to take a tour of possible methods to generally solve a system of linear equations. We will come back to the mobile in Section 5.3.

5.2   Built-In MATLAB Solvers

There are many different methods to find the solution of a system of linear equations; they are usually covered in linear algebra classes. Luckily for us, because of the importance of this type of problem, there are many different ready-made libraries to efficiently attack this problem. Thus, we will skip the internal details of these methods and use MATLAB's built-in functions and operators.

5.2.1   The inverse matrix method

There is an analytical solution for Equation 5.14. Let's multiply the left and right-hand sides of this equation by inverse matrix **A**.

$$\mathbf{A}^{-1}\mathbf{A}\mathbf{x} = \mathbf{A}^{-1}\mathbf{B} \tag{5.17}$$

Since $\mathbf{A}^{-1}\,\mathbf{A}$ equals the identity matrix, it can be omitted.

Analytical solution

$$\mathbf{x} = \mathbf{A}^{-1}\mathbf{B} \tag{5.18}$$

which is applicable only if the determinate of **A** is not equal to zero, that is, $[\det(\mathbf{A}) \neq 0]$.

MATLAB's implementation of the inverse matrix method

```
x=inv(A)*B;
```
(5.19)

There is a price for this straightforward implementation. The calculation of the inverse matrix is computationally taxing.

5.2.2 Solution without inverse matrix calculation

If you ever solved a system of linear equations yourself, you know that there are methods that do not require the inverse matrix, such as Gaussian elimination, which is usually taught in algebra classes. Since the goal of this method is only to get the solution without any sidetracking, its implementation is, usually, much faster.

MATLAB has its own way to get a solution via this route.

MATLAB's way via the left division operator

```
x=A \ B;
```
(5.20)

5.2.3 Which method to use

The left division method is significantly faster then the one shown in Equation 5.19, especially when the size of matrix **A** is larger than 1000×1000. Nevertheless, there are situations when we want to use the inverse matrix method. These are cases when we seek solutions for the same matrix **A** but different vectors **B**. For the case of the mobile problem, it is when we keep the same weights but change the rod lengths from one mobile to another. In this case, we can pre-calculate the inverse matrix once, which takes some time, and then reuse it for the different **B** vectors. The matrix multiplication takes almost no time when compared with the inverse or the left division calculation execution times.

We can demonstrate it with a "synthetic" example. We will generate matrix **A** and vectors **B** with random elements and time the *execution time* with tic and toc commands.

```
Sz = 4000; % matrix dimension
A = rand(Sz, Sz);
B = rand(Sz, 1);
tStart = tic;
x=A \ B;
tElapsed = toc(tStart);
```

The execution time is `tElapsed = 1.25 s.`*

Now let's time the inverse matrix method for the same **A** and **B**.

```
tStart = tic;
invA = inv(A);
x= invA * B;
tElapsed = toc(tStart);
```

In this case, it took `tElapsed = 3.60 s`, which is more than twice as long. Let's see what will happen if we look for a new solution when only **B** changes.

```
B = rand(Sz, 1); % new vector B
tStart = tic;
x= invA * B; % invA is already precalculated
tElapsed = toc(tStart);
```

In this case, `tElapsed = 0.05`, which is more than an order of magnitude faster than any of the original calculations.

5.3   Solution of the Mobile Problem with MATLAB

Now, we are equipped to solve the mobile problem from Section 5.1. We need to assign numerical values to the weights and rod lengths. For example, we can say that $w_1 = 20$, $w_2 = 5$, $w_3 = 3$, $w_4 = 7$, $w_5 = 2$, $w_6 = 3$, $L_{12} = 2$, $L_{34} = 1$, and $L_{56} = 3$. In this case, $w_{26} = 20$ and $w_{46} = 12$. With these definitions, the symbolical matrix in Equation 5.16 becomes

$$
\begin{pmatrix}
20 & -20 & 0 & 0 & 0 & 0 \\
0 & 0 & 3 & -12 & 0 & 0 \\
0 & 0 & 0 & 0 & 2 & -3 \\
1 & 1 & 0 & 0 & 0 & 0 \\
0 & 0 & 1 & 1 & 0 & 0 \\
0 & 0 & 0 & 0 & 1 & 1
\end{pmatrix}
\begin{pmatrix}
x_1 \\ x_2 \\ x_3 \\ x_4 \\ x_5 \\ x_6
\end{pmatrix}
=
\begin{pmatrix}
0 \\ 0 \\ 0 \\ 2 \\ 1 \\ 3
\end{pmatrix}
\tag{5.21}
$$

Now, we are ready to program it.

Listing 5.1 `mobile.m` (available at `http://physics.wm.edu/`
`programming_with_MATLAB_book/./ch_functions_and_scripts/code/`
`mobile.m`)

```
A=[ ...
20, -20,  0,   0,  0,   0; ...
 0,   0,  3, -12,  0,   0; ...
 0,   0,  0,   0,  2,  -3; ...
 1,   1,  0,   0,  0,   0; ...
```

---

* Your execution time will be different, since it is hardware dependent. However, the ratios between this and the following elapsed times will be about the same.

```
   0,     0,   1,    1,   0,    0; ...
   0,     0,   0,    0,   1,    1; ...
   ]
B= [ 0; 0; 0; 2; 1; 3 ]
% 1st method
x=inv(A)*B
% 2nd method
x=A\B
```

The answer is the same in both cases:

```
x =
   1.0000
   1.0000
   0.8000
   0.2000
   1.8000
   1.2000
```

5.3.1 Solution check

It is good idea to check the calculations. To do this, we rearrange Equation 5.14 as

$$\mathbf{A}\mathbf{x} - \mathbf{B} = \mathbf{0} \tag{5.22}$$

where $\mathbf{0}$ is a vector of zeros.

We perform the check

```
>> A*x-B
   1.0e-15 *
      0
      0
      0
      0
      0.2220
      0
```

We expected all zeros, but some elements of the resulting vector are non-zero. Is this a sign of an error? Not really, in this case. We should recall material about round-off errors (see Section 1.5). The deviations from zero are many orders of magnitude smaller than a typical value of the $\mathbf{A}$ or $\mathbf{B}$ element, so everything is as expected.

5.4 Example: Wheatstone Bridge Problem

A system of linear equations often arise in electronics when we need to calculate currents flowing in a circuit and voltage drops across components.

In Figure 5.2, we can see the canonical Wheatstone bridge circuit. A common task associated with this circuit is to find its equivalent resistance. If you know a bit of electronics, you might attempt to reduce this circuit to a set of parallel or series connections. Let me assure you: this will not work except in a few special cases.

The proper way to attack this problem is to connect an imaginary battery to the ends of the bridge circuit (see Figure 5.3), calculate what is the current drained from the battery (I_6), and calculate the resistance by application of the Ohm law:

$$R_{eq} = \frac{V_b}{I_6} \tag{5.23}$$

To find all currents, we need to use two Kirchhoff laws: the sum of the currents in and out of a node is zero (see the first three equations in Listing 5.2), and the total voltage drop in a complete loop is zero (the remaining three equations). For known resistor values $R_1, R_2, \ldots, R_5$, and the battery voltage V_b (which we can set to anything we like here), this forms six linear equations for six unknown currents $I_1, I_2, \ldots, I_6$. These equations are spelled out in the annotated Listing 5.2.

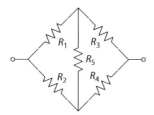

Figure 5.2 The Wheatstone bridge circuit.

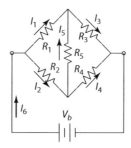

Figure 5.3 The Wheatstone bridge with connected battery.

Listing 5.2 `wheatstone_bridge.m` (available at `http://physics.wm.edu/programming_with_MATLAB_book/./ch_functions_and_scripts/code/wheatstone_bridge.m`)

```
%% Wheatstone bridge calculations
R1=1e3; R2=1e3; R3=2e3; R4=2e3; R5=10e3;
Vb=9;
A=[
   -1, -1,  0,   0,  0,  1;  % I1+I2=I6   eq1
    1,  0, -1,   0,  1,  0;  % I1+I5=I3   eq2
    0,  1,  0,  -1, -1,  0;  % I4+I5=I2   eq3
%   0,  0,  1,   1,  0, -1;  % I3+I4=I6   eq4
% above would make a  linear combination
% of the following  eq1+eq2=-(e3+eq4)
    0,  0, R3, -R4, R5,  0;  % R3*I3+R5*I5=R4*I4 eq4a
   R1,  0, R3,   0,  0,  0;  % R1*I1+R3*I3=Vb    eq5
  -R1, R2,  0,   0, R5,  0   % R2*I2+R5*I5=R1*I1 eq6
]
B=[0; 0; 0; 0; Vb; 0];

% Find currents
I=A\B

% equivalent resistance of the Wheatstone bridge
Req=Vb/I(6)
```

Once we run this script, we will see that Req=1500 Ohms.

There is still a question: did we set the system of equations correctly? It can be shown that $I_6 = 0$ for any R_5, if $R_1/R_2 = R_3/R_4$.* You can set up resistances accordingly in the code in Listing 5.2 to confirm this.

5.5 Self-Study

Problem 5.1

It is possible to draw a parabola through any three points in a plane. Find coefficients a, b, and c for a parabola described as $y = ax^2 + bx + c$, which passes through the points $p_1 = (-10, 10)$, $p_2 = (-2, 12)$, and $p_3 = (12, 10)$. Compare your results with the output of the `polyfit` function.

Problem 5.2

It is possible to draw a fourth degree polynomial through five points in a plane. Find the coefficients of such a polynomial that passes through the following points:

---

* Before inexpensive calibrated multimeters became widespread, it was common to find an unknown resistor (say R_3) by tuning a "programmable" resistor (e.g., R_4) until the I_5 is 0 (this requires only a galvanometer). If R_1 and R_2 are known then $R_3 = R_4 R_1/R_2$. This is why the Wheatstone bridge is such an important circuit.

$p_1 = (0,0)$, $p_2 = (1,0)$, $p_3 = (2,0)$, $p_4 = (3,1)$, and $p_5 = (4,0)$. Compare your results with the output of the `polyfit` function.

Problem 5.3

Set values of resistors such that $R_1/R_2 = R_3/R_4$, and confirm that $I_5 = 0$ for any R_5.

CHAPTER 6

Fitting and Data Reduction

There are many cases when fitting and data reduction are necessary, and MATLAB is useful in solving such problems. This chapter starts by defining fitting and providing a worked example. We then discuss parameter uncertainty estimations and how to evaluate and optimize the resulting fit.

6.1 Necessity for Data Reduction and Fitting

Modern day experiments generate large amounts of data, but humans generally cannot operate simultaneously with more than a handful of parameters or ideas. As a result, very large, raw datasets become virtually useless unless there are efficient, consistent, and reliable methods to reduce the data to a more manageable size. Moreover, the sciences are inductive in nature and ultimately seek to define formulas and equations that simulate or model the reality provided to us in raw data, making data reduction and fitting a crucial aspect of scientific research and inquiry.

In the old days when computers were not available, the most common data reduction methods were calculations of the mean and the standard deviations of a data sample. These methods are still very popular. However, their predictive and modeling powers are very limited.

Ideally, one would like to construct a model that describes (fits) the data samples with only a few free (to modify) parameters. Well-established models that are proved to be true by many generations of scientists are promoted to the status of laws of nature.

Words of wisdom

We should remember that there are no *laws*; there are only hypotheses that explain present observations and have not yet been proved wrong by new data in unexplored regions. Laws of classical physics were replaced by quantum mechanics once enough evidence was collected to highlight discrepancies between models and experiments.

The fitting is the procedure that finds the best values of the free parameters. The process of the model selection is outside of the domain of the fitting algorithm. From a scientific point of view, the model is actually the most important part of the data reduction procedure.

The outcome of the fitting procedure is the set of important parameter values as well as the ability to judge whether the selected model is a good one.

6.2   Formal Definition for Fitting

Suppose someone measured the dependence of an experimental parameter set $\vec{y}$ on another parameter's set $\vec{x}$. We want to extract the unknown model parameters $p_1, p_2, p_3, \ldots = \vec{p}$ via fitting (i.e., finding the best $\vec{p}$) of the model function that depends on $\vec{x}$ and $\vec{p}$: $f(\vec{x}, \vec{p})$. In general, $\vec{x}$ and $\vec{y}$ could be vectors, that is, multidimensional.

Example

- $\vec{x}$ dimensions are the speed of a car and the weight of its load;

- $\vec{y}$ components are the car fuel consumption, the engine temperature, and time to the next repair.

For simplicity, **we will focus our discussion on the one-dimensional case** by dropping the vector notation for x and y.

6.2.1   Goodness of the fit

Have a look at Figure 6.1. We can see the data points (x_i, y_i) and a line corresponding to the functional dependence of our model on a given set of fit parameter values, that is, the fit line. The important points of this fit line are the ones calculated at x_i: $y_{f_i} = f(x_i, \vec{p})$.

We need to define a formal way to estimate the goodness of the fit. One of the well-established methods is the χ^2 (chi squared) test.

χ^2 test

$$\chi^2 = \sum_i (y_i - y_{f_i})^2$$

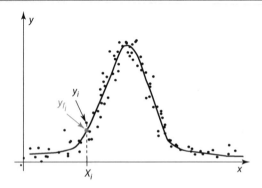

Figure 6.1   Data points and the fit line.

Differences of $(y_i - y_{f_i})$ are called *residuals*. Thus, χ^2 is essentially summed squared distances between the data and the corresponding locations of fit curve points.

For a given set of $\{(x_i, y_i)\}$ and the model function f, the goodness of the fit χ^2 depends only on parameters vector $\vec{p}$ of the model/fit function. The job of the fitting algorithm is simple: find the optimal parameters set $\vec{p}$ that minimizes χ^2 using any suitable algorithm, that is, perform the so-called *least square fit*. Consequently, fitting algorithms are a subclass of problem optimization algorithms (discussed in Chapter 13).

Luckily, we do not have to worry about the implementation of the fitting algorithm, because MATLAB has the fitting toolbox to do the job. The most useful function in this toolbox is `fit`, which governs the majority of work. The fitting functions work faster than general optimization algorithms, since they can specialize in the optimization of a particular quadratic functional dependence of χ^2.

6.3   Fitting Example

The material so far seems to be quite dry, so let's see fitting at work. Suppose we have a set of data points stored in the data file `'data_to_fit.dat'`.* and depicted in Figure 6.2. It looks like data from a resonance contour response or a typical spectroscopic experiment.

Our first priority is to choose a model that might describe the data. Yet again, this is not a problem for the computer, but a problem for the person in charge of

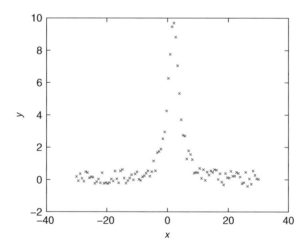

Figure 6.2   Data points.

---

* The file is available at http://physics.wm.edu/programming_with_MATLAB_book/./
  ch_fitting/data/data_to_fit.dat

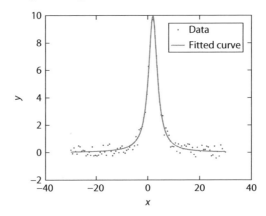

Figure 6.3   Data points and the Lorentzian fit line.

the data analysis. For this example, we choose the Lorentzian shape to fit the data:

$$y = \frac{A}{1 + \left(\frac{x-x_0}{\gamma}\right)^2} \tag{6.1}$$

where:

A is the amplitude (height) of the peak

x_0 is the position of the peak

γ is the peak half width at half maximum level

We decide not to search for an additional background or offset term, since data y-values seem to be around zero, far away from the resonance. Overall, we have three free parameters to fit the data: A, x_0, and γ. The knowledge of these three parameters is sufficient to describe the experimental data, which results in the amount of data needing storage being drastically reduced.

Listing 6.1 demonstrates the fitting procedure.

Listing 6.1 `Lorentzian_fit_example.m` (available at `http://physics.wm.edu/programming_with_MATLAB_book/./ch_fitting/code/Lorentzian_fit_example.m`)

```
%% load initial data file
data=load('data_to_fit.dat');
x=data(:,1); % 1st column is x
y=data(:,2); % 2nd column is y

%% define the fitting function with fittype
% notice that it is quite human readable
% Matlab automatically treats x as independent variable
f=fittype(@(A,x0,gamma, x) A ./ (1 +((x-x0)/gamma).^2) );
```

```
%% assign initial guessed parameters
% [A, x0, gamma] they are in the order of the appearance
% in the above fit function definition
pin=[3,3,1];

%% Finally, we are ready to fit our data
[fitobject] = fit (x,y, f, 'StartPoint', pin)
```

The fitting is actually happening in the last line, where the `fit` function is executed. Everything else is the preparation of the model and data for the `fit` function. The results of the fitting, that is, the values of the free parameters, are part of the `fitobject`. Let's have a look at them and their confidence bounds.

```
fitobject =
     General model:
     fitobject(x) = A./(1+((x-x0)/gamma).^2)
     Coefficients (with 95% confidence bounds):
       A =          9.944  (9.606, 10.28)
       x0 =         1.994  (1.924, 2.063)
       gamma =       2.035  (1.937, 2.133)
```

It is a good idea to visually check the quality of the fit, so we execute

```
%% Let's see how well our fit follows the data
plot(fitobject, x,y, 'fit')
set(gca,'FontSize',24); % adjusting font size
xlabel('x');
ylabel('y');
```

The resulting plot of the data and the fit line are depicted in Figure 6.3.

6.4 Parameter Uncertainty Estimations

How would one estimate the confidence bounds? Well, the details of the MATLAB algorithm are hidden from us, so we do not know. But one of the possible ways would be the following. Suppose we find the best set of free parameters $\vec{p}$ that results in the smallest possible value of χ^2; then, the uncertainty of the ith parameter (Δp_i) can be estimated from the following equation:

$$\chi^2(p_1, p_2, p_3, \ldots p_i + \Delta p_i, \ldots) = 2\chi^2(p_1, p_2, p_3, \ldots p_i, \ldots) = 2\chi^2(\vec{p}) \qquad (6.2)$$

This is a somewhat simplistic view of this issue, since the free parameters are often coupled. The proper treatment of this problem is discussed in [4].

6.5   Evaluation of the Resulting Fit

The visualization of the fit line over the data is a natural step for the fit quality assessment, and it should not be skipped, but we need a more formal set of rules.

Good fits should have the following properties:

1. The fit should use the smallest possible set of fitting parameters.
   - With enough fitting parameters you can ~~fit an elephant through the eye of a needle~~ make a fit with zero residuals, but this is unphysical, since *the experimental data always have uncertainties* in the measurements.
2. The residuals should be randomly scattered around 0 and have no visible trends.
3. The root mean square of residuals $\sigma = \sqrt{\frac{1}{N}\sum_i^N (y_i - y_{f_i})^2}$ should be similar to the experimental uncertainty (Δy) for y.

Words of wisdom

Condition 3 is often overlooked, but you should keep your eyes on it. If $\sigma \ll \Delta y$, you are probably over-fitting, that is, trying to extract from the data what is not there. Alternatively, you do not know the uncertainties of your apparatus, which is even more dangerous. The σ also can give you an estimate of the experimental error bars if they are unaccessible by other means.

---

4. Fits should be robust: the addition of new experimental points must not change the fitted parameters much.

Words of wisdom

Stay away from the higher ordered polynomial fits.

- A line is good, and sometimes a parabola is also good.
- Use anything else only if there is a deep physical reason for it.
- Such higher ordered polynomial fits are usually useless, since every new data point addition tends to drastically modify the fit parameters.

Equipped with these rules, we make the plot of residuals.

```
%% Let's see how well our fit follows the data
plot(fitobject, x,y, 'residuals')
set(gca,'FontSize',24); % adjusting font size
xlabel('x');
ylabel('y');
```

The resulting plot of the residuals is shown in Figure 6.4. One can see that residuals are randomly scattered around zero and have no visible long-term trends. Also, the typical spread of residuals is about 0.5, similar to the data point to point fluctuation, which is especially easy to eyeball on shoulders of the resonance (see Figure 6.2). Thus, at the very least, Conditions 2 and 3 listed in Section 6.5 are true. We also used only three free parameters, so it is unlikely that we can do any better, since we need to provide parameters that govern height, width, and position of resonance. So, condition 1 is also satisfied. The robustness of the fit (condition 4) can be estimated by splitting the data into two sets (e.g., choosing every second point for a given set) and then running the fitting procedure for each of them followed by a comparison of the resulting fit parameters for each of the sets. This exercise is left to the reader.

6.6 How to Find the Optimal Fit

Fitting, in general, is not able to find the best parameter set. It finds only the one that guarantees the local optimal. We will talk more about this in the optimization chapter (see Chapter 13). Such a local optimum might result in an awful fit. As a result, you can often hear people say that the fitting is an art or even witchcraft. As always, what is actually meant is that the person does not understand the rules under which the fitting algorithm operates and has a futile hope that random tweaks in the initial guess will lead to success, that is, that a computer will magically present the proper solution. In this section, we will try to move from the domain of witchcraft to the domain of reliable methods.

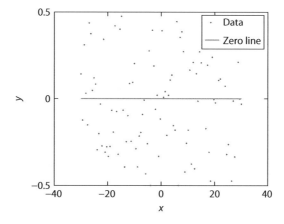

Figure 6.4   Plot of residuals of the Lorentzian fit.

The key to success is in the proper choice of the starting guess.

Words of wisdom

It is very naive to hope that the proper initial guess will be the result of a random selection. The fit algorithm, indeed, performs miraculously well even with a bad starting point. However, one needs to know which fit parameter governs a particular characteristic of the fit line. Otherwise, you can look for a good fit for the rest of your life. After all, computers are only supposed to assist us, not to think for us.

Let me first demonstrate how a bad fit guess leads nowhere. I will modify only one parameter, the initial guess pin=[.1,25,.1], in listing 6.2, which is otherwise identical to Listing 6.1.

Listing 6.2 bad_Lorentzian_fit_example.m (available at http://physics.wm. edu/programming_with_MATLAB_book/./ch_fitting/code/ bad_Lorentzian_fit_example.m)

```
%% load initial data file
data=load('data_to_fit.dat');
x=data(:,1); % 1st column is x
y=data(:,2); % 2nd column is y

%% define the fitting function with fittype
% notice that it is quite human readable
% Matlab automatically treats x as independent variable
f=fittype(@(A,x0,gamma, x) A ./ (1 +((x-x0)/gamma).^2) );

%% assign initial guessed parameters
% [A, x0, gamma] they are in the order of the appearance
% in the above fit function definition
pin=[.1,25,.1]; % <------------- very bad initial guess!

%% Finally, we are ready to fit our data
[fitobject] = fit (x,y, f, 'StartPoint', pin)
```

The resulting fit is shown in Figure 6.5. It is easy to see that the fitted line has no resemblance to the data points overall, except just around $x = 25$, where it passes exactly through two data points. The optimization algorithm was locked in the local maximum, which resulted in the bad fit.

Words of wisdom

Usually, the most critical fit parameters are the ones that govern narrow features in the x-space.

Proper fitting procedure
1. Plot the data.
2. Identify the model/formula that should describe the data. This is outside of the computer's domain.
3. Identify which fit parameter is responsible for a particular fit line feature.
4. Make an *intelligent* guess as to the fit parameters based on this understanding.
5. Plot the fit function with this guess and see whether it is up to your expectations.
6. Refine your guess and repeat the above steps until you get a model function curve reasonably close to the data.
7. Ask the computer to do tedious refinements of your guessed parameters, that is, execute the `fit`.
8. The fit will produce fit parameters with confidence bounds; make sure you like what you see.

As you can see, the most important steps are performed *before* execution of the `fit` command.

6.6.1 Example: Light diffraction on a single slit

The following example should help us see the proper fitting procedure.

6.6.2 Plotting the data

Someone provided us with data that have the sensor response (I values) vs. its position (x values) along the diffraction pattern on a screen of the

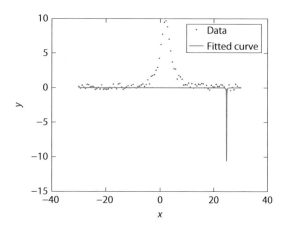

Figure 6.5 Fitting result with the bad starting point.

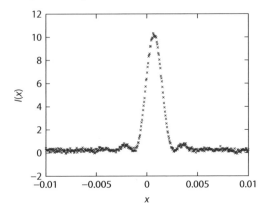

Figure 6.6   Single slit diffraction intensity pattern data.

light passed through a single slit. The plot of the data stored in the file 'single_slit_data.dat'.* is depicted in Figure 6.6.

6.6.3 Choosing the fit model

According to the wave theory of light, the detected intensity of light is given by

$$I(x) = I_0 \left(\frac{\sin\left(\frac{\pi d}{l\lambda}(x - x_0)\right)}{\frac{\pi d}{l\lambda}(x - x_0)} \right)^2 \tag{6.3}$$

where:

d is the slit width
l is distance to the screen
λ is the light wavelength
x_0 is the position of the maximum

This, however, assumes that our detector is ideal, and there is no background illumination. In reality, we need to take into account the offset due to such background (B), which modifies our equation to a more realistic form:

$$I(x) = I_0 \left(\frac{\sin\left(\frac{\pi d}{l\lambda}(x - x_0)\right)}{\frac{\pi d}{l\lambda}(x - x_0)} \right)^2 + B \tag{6.4}$$

The first obstacle for the fitting of our model is in the term

$$\alpha = \frac{\pi d}{l\lambda} \tag{6.5}$$

---

* The file is available at http://physics.wm.edu/programming_with_MATLAB_book/./ch_fitting/data/single_slit_data.dat

d, l, and λ are linked, that is, if someone increases d by 2, we would be able to get the same values of $I(x)$ by increasing either l or λ by 2 to maintain the same ratio. Thus, no fit algorithm will ever be able to decouple these three parameters from the provided data alone. Luckily, the experimenters, who knew about such things, did a good job and provided us with values for $l = .5$ m and $\lambda = 800 \times 10^{-9}$ m. So, by learning the resulting fit parameters, we would be able to tell the slit size d. For now, let's express our formula for intensity with α. The formula looks simpler and requires fewer fit parameters.

$$I(x) = I_0 \left(\frac{\sin (\alpha(x - x_0))}{\alpha(x - x_0)} \right)^2 + B \tag{6.6}$$

We offload its calculation to the function single_slit_diffraction with Listing 6.3.

Listing 6.3 single_slit_diffraction.m (available at http://physics.wm. edu/programming_with_MATLAB_book/./ch_fitting/code/ single_slit_diffraction.m)

```
function [I] = single_slit_diffraction(I0, alpha, B, x0, x
    )
    % calculates single slit diffraction intensity pattern
        on a screen
    % I0 - intensity of the maximum
    % B - background level
    % alpha - (pi*d)/(lambda*l), where
    %   d - slit width
    %   lambda - light wavelength
    %   l - distance between the slit and the screen
    % x - distance across the screen
    % x0 - position of the intensity maximum

    xp = alpha*(x-x0);
    I = I0 * ( sin(xp) ./ xp ).^2  + B;
end
```

6.6.4 Making an initial guess for the fit parameters

Now, let's work on an initial guess for the fit parameters. The B value is probably the simplest; we can see from Equation 6.6 that as x goes to very large values, the first oscillatory term drops (it is $\sim 1/x^2$), and the equation is dominated by B. On other hand, we see that the far edges on Figure 6.6 are above 0 but below 1, so we assign some value in between B_g=0.5 as an initial guess. According to Equation 6.6 definitions, I_0 is the maximum of intensity, disregarding the small B contribution, so we set the I_0 guess as I0_g = 10. Similarly, x_0

is the position of the maximum, so our guess for x_0 (x0_g=.5e−3) seems to be a reasonable value, since the peak is to the right of 0 but to the left of 1e−3. The α value is the trickiest one. First, we recognize that the squared expression in the parentheses of Equation 6.6 is the expression for the *sinc* function. This function is oscillating because of the sin in the numerator, and the amplitude of the oscillation is decreasing with the growth of x, since it sits in the denominator. Importantly, its first location for crossing the background line (x_b) is at the point where $\sin(\alpha(x_b − x_0)) = 0$. So, $\alpha(x_b − x_0) = \pi$. Looking at Figure 6.6, we eyeball that $x_b \approx 0.002$. Thus, a good guess for α is alpha_g = pi/(2e−3 − x0_g).

6.6.5 Plotting data and the model based on the initial guess

We are ready to see whether our intelligent guess is any good. Let's plot the model function with our initial guess values.

Listing 6.4 plot_single_slit_first_guess_and_data.m (available at http:// physics.wm.edu/programming_with_MATLAB_book/./ch_fitting/code/ plot_single_slit_first_guess_and_data.m)

```
% load initial data file
data=load('single_slit_data.dat');
x=data(:,1); % 1st column is x
y=data(:,2); % 2nd column is y

% _g is for guessed parameters
B_g=0.5;
I0_g=10;
x0_g=.5e-3;
alpha_g = pi/(2e-3 - x0_g);

% we have a liberty to choose x points for the model line
Nx= 1000;
xmodel = linspace(-1e-2, 1e-2, Nx);
ymodel = single_slit_diffraction( I0_g, alpha_g, B_g, x0_g
    , xmodel);
plot(x,y,'bx', xmodel, ymodel, 'r-');
legend({'data', 'first guess'});
set(gca,'FontSize',24);
xlabel('x');
ylabel('I(x)');
```

The result is shown in Figure 6.7. It is not a perfect match but pretty close.

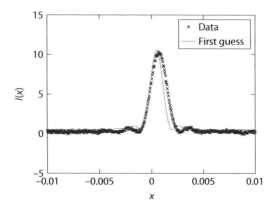

Figure 6.7 Single slit diffraction intensity pattern data and initial guessed fit.

6.6.6 Fitting the data

Now, after all of *our* hard work, we are ready to ask MATLAB to do the fitting.

Listing 6.5 fit_single_slit_data.m (available at http://physics.wm.edu/ programming_with_MATLAB_book/./ch_fitting/code/ fit_single_slit_data.m)

```
% load initial data file
data=load('single_slit_data.dat');
x=data(:,1); % 1st column is x
y=data(:,2); % 2nd column is y

%% defining fit model
f=fittype(@(I0, alpha, B, x0, x) single_slit_diffraction(
    I0, alpha, B, x0, x));

%% prepare the initial guess
% _g is for guessed parameters
B_g=0.5;
I0_g=10;
x0_g=.5e-3;
alpha_g = pi/(2e-3 - x0_g);

% pin = [I0, alpha, B, x0] in order of appearance in
    fittype
pin = [ I0_g, alpha_g, B_g, x0_g ];

%% Finally, we are ready to fit our data
[fitobject] = fit (x,y, f, 'StartPoint', pin)
```

The `fitobject` is listed below:

```
fitobject =
    General model:
    fitobject(x) = single_slit_diffraction(I0,alpha,B,x0,x)
    Coefficients (with 95% confidence bounds):
      I0 =         9.999  (9.953, 10.04)
      alpha =        1572  (1565, 1579)
      B =         0.1995  (0.1885, 0.2104)
      x0 =     0.0006987  (0.0006948, 0.0007025)
```

The resulting fit and its residuals are shown in Figure 6.8, and are generated with

Listing 6.6 `plot_fit_single_slit_data.m` (available at `http://physics.wm.edu/programming_with_MATLAB_book/./ch_fitting/code/plot_fit_single_slit_data.m`)

```
%% plot the data, resulting fit, and residuals
plot(fitobject, x,y, 'fit','residuals')
xlabel('x');
ylabel('y');
```

The residuals are scattered around zero, which is the sign of a good fit.

6.6.7 Evaluating uncertainties for the fit parameters

The `fitobject` provides parameters' values and confidence bounds, so we can estimate the fit parameters' uncertainties or error bars. We can calculate the width of the slit $d = \alpha l \lambda / \pi$ and its uncertainties as follows:

```
%% assigning values known from the experiment
l = 0.5; % the distance to the screen
lambda = 800e-9; % the wavelength in m

%% reading the fit results
alpha = fitobject.alpha; % note .alpha, fitobject is an
    object
ci = confint(fitobject); % confidence intervals for all
    parameters
alpha_col=2; % recall the order of appearance  [I0, alpha,
    B, x0]
dalpha = (ci(2, alpha_col) - ci(1,alpha_col))/2; %
    uncertainty of alpha

%% the width related calculations
a = alpha*l*lambda/pi; % the slit width estimate
da = dalpha*l*lambda/pi; % the slit width uncertainty
```

```
a=2.0016e-04
da=9.2565e-07
```

In particular for the slit width, MATLAB provides way too many digits; they make no sense with the above estimation of the uncertainties (da). We have to do the *proper rounding* ourselves; the slit width is $a = (2.002 \pm 0.009) \times 10^{-4}$ m. Our error bar is less than 0.5%, which is quite good.

Words of wisdom

Take the fit uncertainties with a ~~grain~~ pound of salt. After all, the algorithm has no knowledge about the experiment. It is possible that your data set was very favorable for the particular value of a given free parameter. Run your experiment again several times and see new estimates of the free parameters, compare with the old values, and then decide about parameters uncertainties.

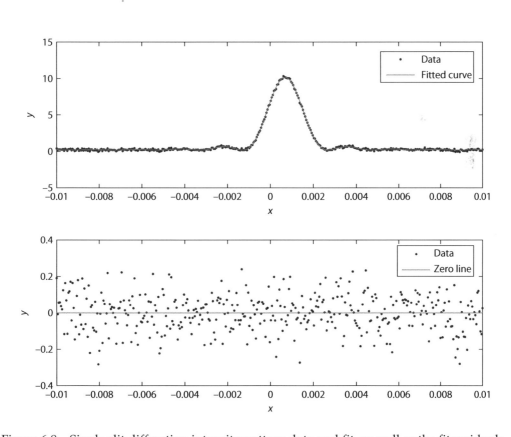

Figure 6.8 Single slit diffraction intensity pattern data and fit, as well as the fit residuals.

6.7 Self-Study

- Review the function handle operator @: use `help function_handle`.

- Pay attention to error bars/uncertainties; **report them**.

- Use built-in `fittype` to define a fitting function with the following call to `fit` to do the fitting.

Problem 6.1

Recall one of the problems from Section 4.6.

Download data file.`'hwOhmLaw.dat'`.[*] It represents the result of someone's attempt to find the resistance of a sample via measuring voltage drop (V), which is the data in the first column, and current (I), listed in the second column, passing through the resistor. Judging by the number of samples, it was an automated measurement.

Using Ohm's law $V = RI$ and a linear fit of the data with one free parameter (R), find the resistance (R) of this sample. What are the error bars/uncertainty of this estimate? Does it come close to the one that you obtained via the method used in Section 4.6? Do not use the fitting menu available via the graphical user interface; use a script or a function to do it.

Problem 6.2

You are making a speed detector based on the Doppler effect. Your device detects dependence of the signal strength vs. time, which is recorded in the data file. `'fit_cos_problem.dat'`.[†] The first column is time, and the second is the signal strength.

Fit the data with

$$A \cos(\omega t + \phi)$$

where:
A, ω and
   ϕ are the amplitude, the frequency and the phase of the signal
   t is time

Find fit parameters (the amplitude, the frequency, and the phase of the signal) and their uncertainties.

Problem 6.3

This is for the physicists among us. Provided that the radar in problem 2 was using radio frequency, could you estimate the velocity measurement uncertainty? Is it a good detector to measure a car's velocity?

---

[*] The file is available at http://physics.wm.edu/programming_with_MATLAB_book/./ch_fitting/data/hwOhmLaw.dat

[†] The file is available at http://physics.wm.edu/programming_with_MATLAB_book/./ch_fitting/data/fit_cos_problem.dat

Problem 6.4

Here is an experiment to do at home. Make a pendulum of variable length (0.1 m, 0.2 m, 0.3 m, and so on up to 1 m). Measure how many round trip (back and forth) swings the pendulum does in 20 s with each particular length (clearly, you will have to round to the nearest integer). Save your observations into a simple text file with "tab" separated columns. The first column should be the length of the pendulum in meters, and the second column should be the number of full swings in 20 s.

Write a script that loads this data file and extract acceleration due to gravity (g) from the properly fitted experimental data. Recall that the period of the oscillation of a pendulum with the length L is given by the following formula:

$$T = 2\pi\sqrt{\frac{L}{g}}$$

Problem 6.5

In optics, the propagation of laser beams is often described in the Gaussian beam's formalism. Among other things, it says that the optical beam intensity cross section is described by the Gaussian profile (hence, the name of the beams).

$$I(x) = A\exp\left(-\frac{(x-x_0)^2}{w^2}\right) + B$$

where:

    A is the amplitude
    x_0 is the position of the maximum intensity
    w is the characteristic width of the beam (width at $1/e$ intensity level)
    B is the background illumination of the sensor

Extract the A, x_0, w, and B with their uncertainties from the real experimental data contained in the file, 'gaussian_beam.dat'.[*] where the first column is the position (x) in meters, and the second column is the beam intensity in arbitrary units.

Does the suggested model describe the experimental data well? Why?

Problem 6.6

Fit the data from the file 'data_to_fit_with_Lorenz.dat'.[†] with the Gaussian profile in problem 5. Is the resulting fit good or not? Why? Compare it with the Lorentzian model (see Equation 6.1).

---

[*] The file is available at http://physics.wm.edu/programming_with_MATLAB_book/./ ch_fitting/data/gaussian_beam.dat
[†] The file is available at http://physics.wm.edu/programming_with_MATLAB_book/./ ch_fitting/data/data_to_fit_with_Lorenz.dat

CHAPTER 7

Numerical Derivatives

In this chapter we discuss methods for finding numerical derivatives of functions. We discuss forward, backward, and central difference methods to estimate the derivative as well as ways to estimate their algorithmic errors. We show that the central difference method is superior to others.

The derivative of a function is a slope of a tangent line to the plot of the function at a given input value (e.g., Figure 7.1 case (a)). There is often a need to calculate the derivative of a function, such as when using the Newton–Raphson root finding algorithm from Section 8.7. The function could be quite complex, and spending effort to derive the analytical expression for the derivative yourself may not be ideal. Alternatively, the function implementation might not even be available to us. For these cases, we resort to a numerical estimate of the function derivative.

7.1 Estimate of the Derivative via the Forward Difference

We might look at the mathematical definition of the derivative

$$f'(x) = \lim_{h \to 0} \frac{f(x+h) - f(x)}{h} \tag{7.1}$$

and implement the numerical estimate ($f'_c(x)$) of the derivative via the *forward difference*

$$f'_c(x) = \frac{f(x+h) - f(x)}{h} \tag{7.2}$$

which essentially approximates the derivative with the finite step h (see Figure 7.1 case (b)). The MATLAB implementation is shown in Listing 7.1.

Listing 7.1 `forward_derivative.m` (available at `http://physics.wm.edu/` `programming_with_MATLAB_book/./ch_derivatives/code/` `forward_derivative.m`)

```
function dfdx = forward_derivative( f, x, h )
% Returns derivative of the function f at position x
% f is the handle to a function
% h is step, keep it small
dfdx = (f(x+h) - f(x))/h;
end
```

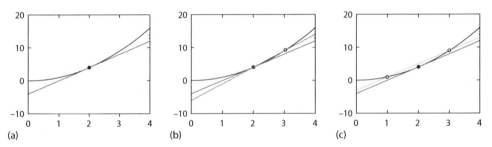

Figure 7.1   (a) The plot of the $f(x) = x^2$ function and its tangent line at the point $(x = 1, y = 1)$. The comparison between the tangent line calculated analytically, as is shown in (a), and by the difference methods: the forward difference (b) and the central difference methods (c). The step h is equal to 1.

Let's check this implementation with $f(x) = x^2$. The derivative of this is $f'(x) = 2x$. So, we expect to see $f'(1) = 2$. At first, we calculate the derivative at x = 1 with h = 1e−5:

```
>> f = @(x) x.^2';
>> forward_derivative(f, 1, 1e-5)
ans =
    2.0000
```

It is very tempting to make h as small as possible to mimic the limit in the mathematical definition. We decrease h:

```
>> forward_derivative(f, 1, 1e-11)
ans =
    2.0000
```

So far, so good; the result is still correct. But if we decrease h further, we get

```
>> forward_derivative(f, 1, 1e-14)
ans =
    1.9984
```

which is imprecise. Somewhat surprisingly, if we make h even smaller, we get

```
>> forward_derivative(f, 1, 1e-15)
ans =
    2.2204
```

This deviates from the correct result even further. This is due to round-off errors (see Section 1.5) and they will get worse as h gets smaller.

7.2 Algorithmic Error Estimate for Numerical Derivative

Unfortunately, there is more to worry about than just round-off errors. Let's assume for a second that the round-off errors are not a concern and a computer can do the calculation precisely. We would like to know what kind of error we are facing by using Equation 7.2. Recall that according to the Taylor series,

$$f(x+h) = f(x) + \frac{f'(x)}{1!}h + \frac{f''(x)}{2!}h^2 + \cdots \tag{7.3}$$

So, we can see that the computed approximation of derivative (f'_c) calculated via the forward difference approximates the true derivative f' as

$$f'_c(x) = \frac{f(x+h) - f(x)}{h} = f'(x) + \frac{f''(x)}{2}h + \cdots$$

From this equation, we can see to the first order of h.

Algorithm error for the forward difference

$$\varepsilon_{fd} \approx \frac{f''(x)}{2}h \tag{7.4}$$

This is quite bad, since the error is proportional to h.

Example

Let's consider the function

$$f(x) = a + bx^2$$
$$f(x+h) = a + b(x+h)^2 = a + bx^2 + 2bxh + bh^2$$
$$f'_c(x) = \frac{f(x+h) - f(x)}{h} \approx 2bx + bh$$

For small $x < b/2$, the algorithm error dominates the derivative approximation.

For a given function and a point of interest, there is an optimal value of h, where both the round-off and the algorithm errors are small and about the same.

7.3   Estimate of the Derivative via the Central Difference

Let's now estimate the derivative via the average of the forward and backward difference.

$$f'_c(x) = \frac{1}{2}\left(\frac{f(x+h) - f(x)}{h} + \frac{f(x) - f(x-h)}{h}\right) \tag{7.5}$$

We can see that the *backward difference* (the second term above) is identical to the forward formula once we plug $-h$ into Equation 7.2 and flip the overall sign. With trivial simplification, we get the *central difference* expression.

The central difference estimate of the derivative

$$f'_c(x) = \frac{f(x+h) - f(x-h)}{2h} \tag{7.6}$$

We probably would not expect any improvement, since we are combining two methods that both have algorithmic errors proportional to h. However, the errors come with different signs and thus cancel each other. We need to follow the Taylor series up to the term proportional to the f''' to calculate the algorithmic error.

Algorithm error of the central difference derivative estimate

$$\varepsilon_{cd} \approx \frac{f'''(x)}{6} h^2 \tag{7.7}$$

The error is quadratic to h, which is a significant improvement (compare cases (b) and (c) in Figure 7.1).

Example

Using the same function as in the previous example

$$f(x) = a + bx^2$$
$$f(x+h) = a + b(x+h)^2 = a + bx^2 + 2bxh + bh^2$$
$$f(x-h) = a + b(x-h)^2 = a + bx^2 - 2bxh + bh^2$$
$$f'_c(x) = \frac{f(x+h) - f(x-h)}{2h} = 2bx$$

This is the exact answer, which is not very surprising, since all derivatives with an order higher than 3 are zero. The algorithmic error in this case is zero.

We get a much better derivative estimate for the same computational price: we still have to evaluate our function only twice to get the derivative. Thus, the central difference should be used whenever possible, unless we need to reduce computational load by reusing some values of the function calculated at prior steps.[*]

---

[*] In some calculations, a single evaluation of a function can take days or even months.

7.4   Self-Study

Problem 7.1

Plot the $\log_{10}$ of the absolute error (when compared with the true value) of the $\sin(x)$ derivative at $x = \pi/4$ calculated with forward and central difference methods vs. the $\log_{10}$ of the step size h value. See `loglog` help for plotting with the logarithmic axes. The values of h should cover the range $10^{-16} \cdots 10^{-1}$ (read about MATLAB's `logspace` function designed for such cases).

Do the errors scale as predicted by Equations 7.4 and 7.7?

Why does the error decrease with h and then start to increase?

Note: the location of the minimum of the absolute error indicates the optimal value of h for this particular case.

CHAPTER 8

Root Finding Algorithms

In this chapter, we cover root finding algorithms. We present the general strategy and several classic algorithms: bisection, false position, Newton–Raphson, and Ridders. Then, we discuss potential pitfalls of numerical methods and review advantages and disadvantages of the classic algorithms. We also show how MATLAB's built-in methods are used to find the root of an equation.

8.1 Root Finding Problem

We will discuss several general algorithms that find a solution of the following canonical equation:

$$f(x) = 0 \tag{8.1}$$

Here, $f(x)$ is any function of one scalar variable x,[*] and an x that satisfies Equation (8.1) is called a *root* of the function f. Often, we are given a problem that looks slightly different:

$$h(x) = g(x) \tag{8.2}$$

But it is easy to transform it to the canonical form with the following relabeling:

$$f(x) = h(x) - g(x) = 0 \tag{8.3}$$

Example

$$3x^3 + 2 = \sin x \quad \rightarrow \quad 3x^3 + 2 - \sin x = 0 \tag{8.4}$$

For some problems of this type, there are methods to find analytical or closed-form solutions. Recall, for example, the quadratic equation problem, which we

---

[*] Methods to solve a more general equation in the form $\vec{f}(\vec{x}) = 0$ are considered in Chapter 13, which covers optimization.

discussed in detail in Chapter 4. Whenever it is possible, we should use the closed-form solutions. They are usually exact, and their implementations are much faster. However, an analytical solution might not exist for a general equation, that is, our problem is *transcendental*.

Example

The following equation is transcendental:

$$e^x - 10x = 0 \tag{8.5}$$

We will formulate methods that are agnostic to the functional form of Equation 8.1 in the following text.*

8.2   Trial and Error Method

Broadly speaking, all methods presented in this chapter are of the *trial and error* type. One could attempt to obtain the solution by just guessing it, and eventually, one would succeed. Clearly, the probability of success is quite small in every attempt. However, each guess can provide some clues that would point us in the right direction. The main difference between algorithms is in how the next guess is formed.

A general numerical root finding algorithm is the following:

- Make a guess (x_i).
- Make a new **intelligent** guess (x_{i+1}) based on this trial x_i and $f(x_i)$.
- Repeat as long as the required precision on the function closeness to zero

$$|f(x_{i+1})| < \varepsilon_f \tag{8.6}$$

and solution convergence

$$|x_{i+1} - x_i| < \varepsilon_x \tag{8.7}$$

are not reached. The solution convergence check is optional, but it provides estimates on the solution precision.

---

* MATLAB has built-in functions that can solve the root finding problem. However, programming the algorithms outlined in this chapter has great educational value. Also, studying general methods for finding the solution might help you in the future, when you will have to make your own implementation in a programming language that does not have a built-in root finder. Besides, if you know what is under the hood, you can use it more efficiently or avoid misuse.

8.3 Bisection Method

To understand the bisection method, let's consider a simple game: someone thinks of any integer number between 1 and 100, and our job is to guess it.

If there are no clues provided, we would have to probe every possible number. It might take as many as **100** attempts to guess correctly. Now, suppose that after each guess, we are getting a clue: our guess is "high" or "low" when compared with the number in question. Then, we can split the search region in half with every iteration. We would need at most only **7** attempts to get the answer, if we do it right. Clearly, this is much faster.

Example

Let's say the number to find is 58.

1. For the first guess, we choose the middle point in the interval 1–100, which yields the first guess: 50.

2. We hear that our guess is "low," so we will search in the upper half: 50–100. The second guess will be in the middle of 50–100, that is, 75.

3. We hear that this is "high," so we divide 50–75 in half again. For the third guess, we say 63.

4. We hear "high." We split 50–63 in half again, and say 56.

5. We hear "low." So, we split the upper half of the region 56–63, and say 59.

6. We hear "high." and split the low part of 56–59, and say 57.

7. We hear "low." so we make our final matching guess: 58.

In total, we made 7 guesses, which is the worst-case scenario for this strategy.

The shown example outlines the idea of the bisection method: divide the region into two equal halves, and operate on the remaining half. The following *pseudo-code** for this algorithm, which works for any continuous function provided that we *bracketed* the root, that is, we provided two points at which our function has opposite signs.

---

* The pseudo-code is designed for human reading. It omits parts that are essential for a correct computer implementation.

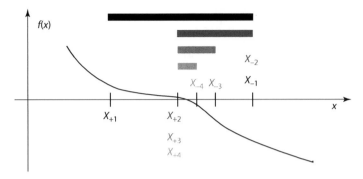

Figure 8.1   The bisection method illustration. $X_{\pm i}$ mark the bracket position on the ith iteration. The root enclosing bracket is indicated by the wide stripe.

Bisection algorithm's pseudo-code

1. Decide on maximal allowed deviation (ε_f) of the function from zero and the root precision (ε_x).

2. Make an initial root enclosing bracket for the search, that is, choose a positive end x_+ and a negative end x_- such that $f(x_+) > 0$ and $f(x_-) < 0$. Note that $+$ and $-$ refer to the function sign and not to the relative position of the bracket ends.

3. Begin the searching loop.

4. Calculate the new guess value $x_g = (x_+ + x_-)/2$.

5. If $|f(x_g)| \leq \varepsilon_f$ and $|x_+ - x_g| \leq \varepsilon_x$, stop: we have found the root with the desired precision.*

6. Otherwise, reassign one of the bracket ends: if $f(x_g) > 0$, then $x_+ = x_g$, else $x_- = x_g$.

7. Repeat the searching loop.

Figure 8.1 shows the first several iterations of the bisection algorithm. It shows with bold stripes the length of the bracketed region. The points marked as $X_{\pm i}$ are positions of the negative ($-$) and positive ($+$) ends of the root enclosing bracket.

The MATLAB implementation of the bisection algorithm is shown in Listing 8.1.

---

* Think about why we are using the modified solution convergence expression and not the condition of Equation 8.7.

Listing 8.1 `bisection.m` (available at `http://physics.wm.edu/` `programming_with_MATLAB_book/./ch_root_finding/code/bisection.m`)

```matlab
function [xg, fg, N_eval] = bisection(f, xn, xp, eps_f, eps_x)
% Solves f(x)=0 with bisection method
%
%  Outputs:
%    xg is the root approximation
%    fg is the function evaluated at final guess f(xg)
%    N_eval is the number of function evaluations
%  Inputs:
%    f is the function handle to the desired function,
%    xn and xp are borders of search, i.e. root brackets,
%    eps_f defines maximum deviation of f(x_sol) from 0,
%    eps_x defines maximum deviation from the true solution
%
%  For simplicity reasons, no checks of input validity are done:
%    it is up to user to check that f(xn)<0 and f(xp)>0,
%    and that all required deviations are positive

%% initialization
xg=(xp+xn)/2; % initial root guess
fg=f(xg);      % initial function evaluation
N_eval=1; % We just evaluated the function

%% here we search for root
while ( (abs(xg-xp) > eps_x) || (abs(fg) > eps_f) )
    if (fg>0)
        xp=xg;
    else
        xn=xg;
    end
    xg=(xp+xn)/2;      % update the guessed x value
    fg=f(xg);          % evaluate the function at xg
    N_eval=N_eval+1; % update evaluation counter
end

%% solution is ready
end
```

An interesting exercise for a reader is to see that the *while* condition is equivalent to the one presented in step 5 of the bisection's pseudo-code. Also, note the use of the *short-circuiting or* operator represented as ||. Please have a look at the MATLAB manual to learn what it does.

8.3.1   Bisection use example and test case

8.3.1.1   Test the bisection algorithm

For practice, let's find the roots of the following equation:

$$(x - 10) \times (x - 20) \times (x + 3) = 0 \tag{8.8}$$

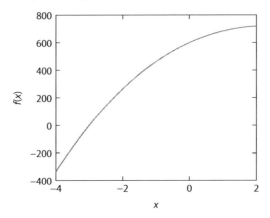

Figure 8.2   Plot of the function to solve $f(x) = (x - 10) \times (x - 20) \times (x + 3)$ in the range from −4 to 2.

Of course, we do not need a fancy computer algorithm to find the solutions: 10, 20, and −3, but knowing the roots in advance allows us to check that we know how to run the code correctly. Also, we will see a typical work flow for the root finding procedure. But most importantly, we can test whether the provided bisection code is working correctly: it is always a good idea to check new code against known scenarios.

We save MATLAB's implementation of the test function into the file ' function_to_solve.m'

Listing 8.2   function_to_solve.m (available at http://physics.wm.edu/ programming_with_MATLAB_book/./ch_root_finding/code/ function_to_solve.m)

```
function ret=function_to_solve(x)
        ret=(x-10).*(x-20).*(x+3);
end
```

It is a good idea to plot the function to eyeball a potential root enclosing bracket. Have a look at the function in the range from −4 to 2 in Figure 8.2. Any point where the function is negative would be a good choice for a negative end, xn = −4 satisfies this requirement. Similarly, we have many choices for the positive end of the bracket where the function is above zero. We choose xp = 2 for our test.

One more thing that we need to decide is the required precision of our solution. The higher the precision, the longer the calculation will run. This is probably not a big factor for our test case. However, we really should ensure that we do not ask for precision beyond the computer's number representation. With 64-bit floats currently used by MATLAB, we definitely should not ask beyond 10^{-16} precision. For this test run, we choose eps_f=1e−6 and eps_x=1e−8.

We find a root of Equation 8.8 with the following code. Notice how we send the handle of the function to solve with @ operator to bisection function.

```
>> eps_x = 1e-8;
>> eps_f = 1e-6;
>> x0 = bisection( @function_to_solve, -4, 2, eps_f, eps_x
    )
x0 = - 3.0000
```

The algorithm seems to yield the exact answer -3. Let's double check that our function is indeed zero at x0.

```
>> function_to_solve(x0)
ans = 3.0631e-07
```

Wow, the answer is not zero. To explain this, we should recall that we see only five significant digits of the solutions, that is, -3.0000; also, with eps_x=1e−6 we requested precision for up to seven significant digits. So, we see the rounded representation of x0 printed out. Once we plug it back into the function, we see that it satisfies our requested precision for the function zero approximation (eps_ f=1e−6) with f(x0) = 3.0631−07 but not much better. The bottom line: we got what we asked for, and everything is as expected.

Notice that we have found only one out of three possible roots. To find the others, we would need to run the bisection function with the appropriate root enclosing brackets for each of the roots. The algorithm itself has no capabilities to choose the brackets.

8.3.1.2   One more example

Now, we are ready to find the root of the transcendental Equation 8.5. We will do it without making a separate file for a MATLAB implementation of the function. Instead, we will use anonymous function f.

```
>> f = @(x) exp(x) - 10*x;
>> x1 = bisection( f, 2, -2, 1e-6, 1e-6)
x1 =
      0.1118
>> [x2, f2, N] = bisection( f, 2, 10, 1e-6, 1e-6)
x2 =
      3.5772
f2 =
    2.4292e-07
N =
      27
```

As we can see, Equation 8.5 has two roots:[*] x1=0.1118 and x2=3.5772. The output of the second call to bisection returns the value of the function at the approximation of the true root, which is f2=2.4292−07 and within the required precision 1e−6. The whole process took only 27 steps or iterations.

---

[*] It is up to the reader to prove that there are no other roots.

8.3.2 Possible improvement of the bisection code

The simplified bisection code is missing validation of input arguments. People make mistakes, typos, and all sorts of misuse. Our `bisection` function has no protection against this. In the example of Section 8.3.1.1, if we accidentally swapped the positive and negative ends of the bracket in the example we have just shown, `bisection` would run forever or at least until we stopped the execution. Try to see such misbehavior by executing

```
>> bisection( @function_to_solve, 2, -4, eps_f, eps_x )
```

Once you are tired of waiting, *interrupt the execution* by pressing Ctrl and c keys together.

> **Words of wisdom**
>
> "If something can go wrong it will."
>
> Murphy's Law.

We should recall good programming practices from Section 4.4 and validate the input arguments.* At the very least, we should make sure that `f(xn)<0` and `f(xp)>0`.

8.4 Algorithm Convergence

We say that the root finding algorithm has the defined convergence if

$$\lim_{k \to \infty} (x_{k+1} - x_0) = c(x_k - x_0)^m \tag{8.9}$$

where:
  x_0 is the true root of the equation
   c is some constant
  m is the *order of convergence*

If for an algorithm $m = 1$, then we say that the algorithm converges linearly. The case of $m > 1$ is called superlinear convergence.

It is quite easy to show (by using the size of the bracket for the upper bound estimate of the distance $x_k - x_0$) that the bisection algorithm has linear rate of convergence ($m = 1$) and $c = 1/2$.

Generally, the speed of the algorithm is related to its convergence order: the higher the better. However, other factors may affect the overall speed. For example, there could be too many intermediate calculations, or the required memory

---

* Never expect that a user will put valid inputs.

size could outgrow the available memory of a computer for an otherwise higher convergence algorithm.

If convergence is known, we can estimate how many iterations of the algorithm are required to reach the required root-reporting precision. Unfortunately, it is often impossible to define convergence order for a general algorithm.

Example

In the bisection method, the initial bracket size (b_0) decreases by a factor of 2 on every iteration. Thus, at step N, the bracket size is $b_N = b_0 \times 2^{-N}$. It should be $< \varepsilon_x$ to reach the required root precision. We need at least the following number of steps to achieve it:

$$N \geq \log_2(b_0/\varepsilon_x) \tag{8.10}$$

Conversely, we are getting an extra digit in the root estimate approximately every 3 iterations.

The bisection method is great: it always works, and it is simple to implement. But its convergence is quite slow. The following several methods attempt to improve the guess by making some assumptions about the function shape.

8.5 False Position (*Regula Falsi*) Method

If the function is smooth and its derivative does not change too much, we can naively approximate our function as a line. We need two points to define a line. We will take negative and positive ends of the bracket as such line-defining points, that is, the line is the chord joining the function bracket points. The point where this chord crosses zero is our new guess point. We will use it to update the appropriate end of the bracket. The *regula falsi* method is illustrated in Figure 8.3.

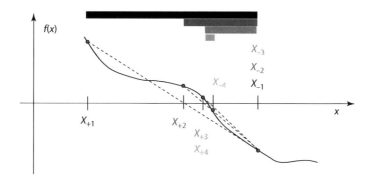

Figure 8.3   The *regula falsi* method illustration. We can see how the new guess is constructed and which points are taken as the bracket ends. A wide stripe indicates the bracket size at a given step.

Since the root remains bracketed all the time, the method is guaranteed to converge to the root value. Unfortunately, the convergence of this method can be slower than the bisection's convergence in some situations, which we show in Section 8.9.

Regula falsi method pseudo-code

1. Choose proper initial bracket for search x_+ and x_- such that $f(x_+) > 0$ and $f(x_-) < 0$.

2. Loop begins.

3. Draw the chord between points $(x_-, f(x_-))$ and $(x_+, f(x_+))$.

4. Make a new guess value at the point of the chord intersection with the x axis:

$$x_g = \frac{x_- f(x_+) - x_+ f(x_-)}{f(x_+) - f(x_-)} \tag{8.11}$$

5. If $|f(x_g)| \leq \varepsilon_f$ and the root convergence is reached

$$(|x_g - x_-| \leq \varepsilon_x) \vee (|x_g - x_+| \leq \varepsilon_x) \tag{8.12}$$

then stop: we have found the solution with the desired precision.

6. Otherwise, update the bracket end: if $f(x_g) > 0$, then $x_+ = x_g$, else $x_- = x_g$.

7. Repeat the loop.

Note: the algorithm resembles the bisection pseudo-code except for the way of updating x_g and checking the x_g convergence.

8.6   Secant Method

The secant method uses the same assumption about the function, that is, it is smooth, and its derivative does not change too widely. Overall, the secant method is very similar to the *regula falsi*, except that we take two arbitrary points to draw a chord. Also, we update the oldest used point for the chord drawing with the newly guessed value, as illustrated in Figure 8.4. Unlike in the false position method, where one of the ends is sometimes (or even never) updated, the ends are always updated, which makes the convergence of the secant algorithm superlinear: the order of convergence m is equal to the *golden ratio* [9], that is, $m = (1 + \sqrt{5})/2 \approx 1.618\ldots$. Unfortunately, because the root is not bracketed, **the convergence is not guaranteed**.

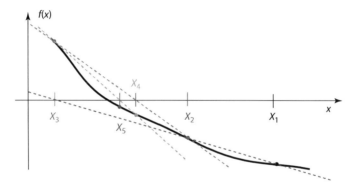

Figure 8.4   The secant method illustration.

Outline of the secant algorithm

1. Choose two arbitrary starting points, x_1 and x_2.

2. Loop begins.

3. Calculate next guess according to the following iterative formula:

$$x_{i+2} = x_{i+1} - f(x_{i+1})\frac{x_{i+1} - x_i}{f(x_{i+1}) - f(x_i)} \tag{8.13}$$

4. Throw away x_i point.

5. Repeat the loop until the required precision is reached.

8.7   Newton–Raphson Method

The Newton–Raphson method also approximates the function with a line. In this case, we draw a line through a guess point $(x_g, f(x_g))$ and make the line's slope equal to the derivative of the function itself. Then, we find where this line crosses the x axis and take this point as the next guess. The process is illustrated in Figure 8.5. The process converges quadratically, that is, $m = 2$, which means that we double the number of significant figures with every iteration [9], although the calculation of the derivative could be as time consuming as calculating the function itself (see e.g., numerical derivative algorithms in Chapter 7), that is, one iteration with Newton–Raphson is equivalent to two iterations with some other algorithm. So, the order of convergence is actually $m = \sqrt{2}$. The downside is that convergence to the root is not guaranteed and is quite sensitive to the starting point choice.

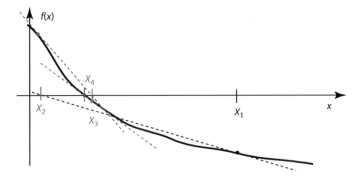

Figure 8.5   The Newton–Raphson method illustration.

Outline of the Newton–Raphson algorithm

1. Choose an arbitrary starting point x_1.

2. Loop begins.

3. Calculate next guess according to the following iterative formula:

$$x_{i+1} = x_i - \frac{f(x_i)}{f'(x_i)} \qquad (8.14)$$

4. Repeat the loop until the required precision is reached.

One has a choice how to calculate f'. It can be done analytically (the preferred method, but it requires additional programming of the separate derivative function) or numerically (for details, see Chapter 7).

In Listing 8.3 you can see the simplified implementation of the Newton–Raphson algorithm without the guess convergence test (this is left as an exercise for the reader).

Listing 8.3 NewtonRaphson.m (available at
http://physics.wm.edu/programming_with_MATLAB_book/./
ch_root_finding/code/NewtonRaphson.m)

```
function [x_sol, f_at_x_sol, N_iterations] = NewtonRaphson(f, xguess, eps_f,
    df_handle)
%    Finds the root of equation f(x)=0 with the Newton-Raphson algorithm
%    f - the function to solve handle
%    xguess - initial guess (starting point)
%    eps_f - desired precision for f(x_sol)
%    df_handle - handle to the derivative of the function f(x)

%  We need to sanitize the inputs but this is skipped for simplicity

        N_iterations=0;  % initialization of the counter
        fg=f(xguess);    % value of function at guess point

        while( (abs(fg)>eps_f) ) % The xguess convergence check is not
            implemented
                xguess=xguess - fg/df_handle(xguess); % evaluate new guess
```

```
                    fg=f(xguess);
                    N_iterations=N_iterations+1;
            end
            x_sol=xguess;
            f_at_x_sol=fg;
end
```

8.7.1 Using Newton–Raphson algorithm with the analytical derivative

Let's see how to call the Newton–Raphson if the analytical derivative is available.
We will solve

$$f(x) = (x - 2) \times (x - 3) \tag{8.15}$$

It is easy to see that the derivative of the above $f(x)$ is

$$f'(x) = 2x - 5 \tag{8.16}$$

To find a root, we should first implement code for f and f':

```
>> f = @(x) (x-2).*(x-3);
>> dfdx = @(x) 2*x - 5;
```

Now, we are ready to execute our call to NewtonRaphson.

```
>> xguess = 5;
>> eps_f=1e-8;
>> [x_1, f_at_x_sol, N_iterations] = NewtonRaphson(f, xguess, eps_f, dfdx)
x1 =
    3.0000
f_at_x_sol =
    5.3721e-12
N_iterations =
    6
```

In only six iterations, we find only the root $x_1 = 3$ out of two possible solutions.
Finding all the roots is not a trivial task. But if we provide a guess closer to another
root, the algorithm will converge to it.

```
>> x2 = NewtonRaphson(f, 1, eps_f, dfdx)
x2 =
    2.0000
```

As expected, we find the second root $x_2 = 2$. Strictly speaking, we find the
approximation of the root, since the x_sol2 is not exactly 2:

```
>> 2-x_sol2
ans =
    2.3283e-10
```

but this is what we expect with numerical algorithms.

8.7.2 Using Newton–Raphson algorithm with the numerical derivative

We will look for a root of the following equation:

$$g(x) = (x - 3) \times (x - 4) \times (x + 23) \times (x - 34) \times \cos(x) \tag{8.17}$$

In this case, we will resort to the numerical derivative, since the $g'(x)$ is too cumbersome. It is likely to make an error during the derivation of the analytical derivative.

First, we implement the general forward difference formula (see why this is not the best method in Section 7.3).

```
>> dfdx = @(x, f, h) (f(x+h)-f(x))/h;
```

Second, we implement $g(x)$.

```
>> g = @(x) (x-3).*(x-4).*(x+23).*(x-34).*cos(x);
```

Now, we are ready to make the **specific** numerical derivative implementation of $g'(x)$. We choose step h=1e−6.

```
>> dgdx = @(x) dfdx(x, g, 1e-6);
```

Finally, we search for a root (pay attention: we use here g and dgdx).

```
xguess = 1.4; eps_f=1e-6; x_sol = NewtonRaphson(g, xguess,
    eps_f, dgdx)
x_sol =
    1.5708
```

Note that $\pi/2 \approx 1.5708$. The algorithm converged to the root that makes $\cos(x) = 0$ and, consequently, $g(x) = 0$.

8.8 Ridders' Method

As the name hints, this method was proposed by Ridders [10]. In this method, we approximate the function from Equation 8.8 with a nonlinear one to take its curvature into account, thus making a better approximation. The trick is in a special form of the approximation function, which is the product of two equations:

$$f(x) = g(x)e^{-C(x-x_r)} \tag{8.18}$$

where:

$g(x) = a + bx$ is a linear function

C is some constant

x_r is an arbitrary reference point.

The advantage of this form is that if $g(x_0) = 0$, then $f(x_0) = 0$, but once we know coefficients a and b, the x where $g(x) = 0$ is trivial to find.

We might expect a faster convergence of this method, since we do a better approximation of the function $f(x)$. But there is a price to pay: the algorithm is a bit more complex, and we need an additional calculation of the function beyond the bracketing points, since we have three unknowns, a, b, and C, to calculate (we have freedom of choice for x_r).

If we choose the additional point location $x_3 = (x_1 + x_2)/2$, then the position of the guess point x_4 is quite straight forward to calculate with a proper bracket points x_1 and x_2.

$$x_4 = x_3 + \text{sign}(f_1 - f_2) \frac{f_3}{\sqrt{f_3^2 - f_1 f_2}} (x_3 - x_1) \tag{8.19}$$

where $f_i = f(x_i)$, and $\text{sign}(x)$ stands for the sign of the function's argument: $\text{sign}(x) = +1$ when $x > 0$, -1 when $x < 0$, and 0 when $x = 0$. The search for the guess point is illustrated in Figure 8.6, where $x_r = x_3$.

Ridders' algorithm outline

1. Find **proper** bracketing points for the root x_1 and x_2. It is irrelevant which is the positive end and which is negative, but the function must have different signs at these points, that is, $f(x_1) \times f(x_2) < 0$.

2. Loop begins.

3. Find the midpoint $x_3 = (x_1 + x_2)/2$.

4. Find a new approximation for the root

$$x_4 = x_3 + \text{sign}(f_1 - f_2) \frac{f_3}{\sqrt{f_3^2 - f_1 f_2}} (x_3 - x_1) \tag{8.20}$$

where

$$f_1 = f(x_1), f_2 = f(x_2), f_3 = f(x_3)$$

5. If x_4 satisfies the required precision and the convergence condition is reached, then stop.

6. Rebracket the root, that is, assign new x_1 and x_2, using old values.
   - One end of the bracket is x_4 and $f_4 = f(x_4)$.
   - The other is whichever of (x_1, x_2, x_3) is closer to x_4 **and provides the proper bracket**.

7. Repeat the loop.

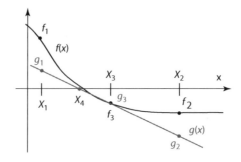

Figure 8.6   The Ridders' method illustration. The reference point position $x_r = x_3$ and $x_3 = (x_1 + x_2)/2$.

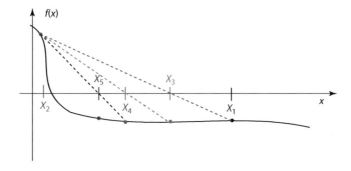

Figure 8.7   The false position slow convergence pitfall. Think what would happen if the long horizontal tail of the function lay even closer to the x-axis.

In Ridders' algorithm, the root is always properly bracketed; thus, the algorithm always converges, and x_4 is always guaranteed to be inside the initial bracket. Overall, the convergence of the algorithm is quadratic per cycle ($m = 2$). However, it requires evaluation of the $f(x)$ twice for f_3 and f_4; thus, it is actually $m = \sqrt{2}$ [10].

8.9   Root Finding Algorithms Gotchas

The root bracketing algorithms are bulletproof and always converge, but convergence of the false position algorithm could be slow. Normally, it outperforms the bisection algorithm, but for some functions, that is not the case. See, for example, the situation depicted in Figure 8.7. In this example, the bracket shrinks by a small amount on every iteration. The convergence would be even worse if the long horizontal tail of the function ran closer to the x axis.

The non-bracketing algorithms, such as Newton–Raphson and secant, usually converge faster than their bracketing counterparts. However, their convergence is not guaranteed! In fact, they may even diverge and run away from the root. Have a look at Figure 8.8, where a pitfall of the Newton–Raphson method is shown. In

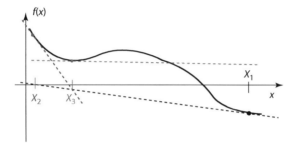

Figure 8.8 The Newton–Raphson method pitfall: the x_4 guess is located far to the right and way farther than original guess x_1.

just three iterations, the guess point moved far away from the root and our initial guess.

> **Words of wisdom**
>
> There is no **silver bullet** algorithm that would work in all possible cases. We should carefully study the function for which the root is searched, and see whether all relevant requirements of an algorithm are satisfied. When unsure, sacrifice speed, and choose a more robust but slower bracketing algorithm.

8.10 Root Finding Algorithms Summary

We have not by any means considered all root finding algorithms. We just covered a small subset. If you are interested in seeing more information, you may read [9].

Here, we present a short summary of root bracketing and non-bracketing algorithms.

Root bracketing algorithms

- Bisection
- False position
- Ridders'

Pros

- Robust that is, always converge

Cons

- Usually slower convergence
- Require initial bracketing

Non-bracketing algorithms

- Newton–Raphson
- Secant

Pros

- Faster
- No need to bracket (just give a **reasonable** starting point)

Cons

- **May not converge**

8.11   MATLAB's Root Finding Built-in Command

MATLAB uses `fzero` to find the root of an equation. Deep inside, `fzero` uses a combination of bisection, secant, and inverse quadratic interpolation methods. The built-in `fzero` has quite a few options, but in the simplest form, we can call it as shown here to solve Equation 8.5.

To search for the root by providing only the starting point,

```
>> f = @(x) exp(x) - 10*x;
>> fzero(f, 10)
ans =
    3.5772
```

We have no control over which of possible roots will be found this way. We can provide the proper bracket within which we want to find a root:

```
>> f = @(x) exp(x) - 10*x;
>> fzero(f, [-2,2])
ans =
    0.1118
```

In this case, the bracket spans from -2 to 2. As you can see, we find the same roots as with our `bisection` implementation discussed in Section 8.3.1.2.

8.12   Self-Study

General requirements:

1. Test your implementation with at least $f(x) = \exp(x) - 5$ and the initial bracket [0,3], but do not limit yourself to only this case.

2. If the initial bracket is not applicable (e.g., in the Newton–Raphson algorithm), use the right end of the test bracket as the starting point of the algorithm.

3. All methods should be tested for the following parameters: eps_f=1e−8 and eps_x=1e−10.

Problem 8.1

Write a proper implementation of the false position algorithm. Define your function as

```
function [x_sol, f_at_x_sol, N_iterations] = regula_falsi(f, xn, xp, eps_f, eps_x)
```

Problem 8.2

Write a proper implementation of the secant algorithm. Define your function as

```
function [x_sol, f_at_x_sol, N_iterations] = secant(f, x1, x2, eps_f, eps_x)
```

Problem 8.3

Write a proper implementation of the Newton–Raphson algorithm. Define your function as

```
function [x_sol, f_at_x_sol, N_iterations] = NewtonRaphson(f, xstart, eps_f, eps_x, df_handle).
```

Note that `df_handle` is a function handle to calculate the derivative of the function f; it could be either an analytical representation of $f'(x)$ or its numerical estimate via the central difference formula.

Problem 8.4

Write a proper implementation of Ridders' algorithm. Define your function as

```
function [x_sol, f_at_x_sol, N_iterations] = Ridders(f, x1, x2, eps_f, eps_x)
```

Problem 8.5

For each of your root finding implementations, find roots of the following two functions:

1. $f1(x) = \cos(x) - x$ with the x initial bracket [0,1]
2. $f2(x) = \tanh(x - \pi)$ with the x initial bracket $[-10, 10]$

Make a comparison table for these algorithms with the following rows:

1. Method name
2. Root of $f1(x)$
3. Initial bracket or starting value used for $f1$
4. Number of iterations to solve $f1$
5. Root of $f2(x)$
6. Initial bracket or starting value used for $f2$
7. Number of iterations to solve $f2$

If an algorithm diverges with the suggested initial bracket, indicate this, appropriately modify the bracket, and show the modified bracket in the above table as well. State your conclusions about the speed and robustness of the methods.

CHAPTER 9

Numerical Integration Methods

In this chapter, we discuss several methods for calculating numerical integrals. We discuss advantages and disadvantages of each method by comparing their algorithmic errors. We cover one-dimensional and multidimensional integrals and potential pitfalls during their calculations.

The ability to calculate integrals is quite important. The author was told that, in the old days, the gun ports were cut into a ship only after it was afloat, loaded with equivalent cargo, and rigged. This is because it was impossible to calculate the water displaced volume, that is, the integral of the hull-describing function, with the rudimentary math known at that time. Consequently, there was no way to properly estimate the buoyant force. Thus, the location of the waterline was unknown until the ship was fully afloat.

Additionally, not every integral can be computed analytically, even for relatively simple functions.

Example

The Gauss error function defined as

$$\mathrm{erf}(y) = \frac{2}{\sqrt{\pi}} \int_0^y e^{-x^2} dx$$

cannot be calculated using only elementary functions.

9.1 Integration Problem Statement

At first, we consider the evaluation of a one-dimensional integral, also called *quadrature*, since this operation is equivalent to finding the area under the curve of a given function.

$$\int_a^b f(x)dx$$

Surprisingly, one does not need to know any high-level math to do so. All you need is a precision scale and scissors. The recipe goes like this: plot the function on some preferably dense and heavy material, cut the area of interest with scissors, measure the mass of the cutout, divide the obtained mass by the density and thickness of the material, and you are done. Well, this is, of course, not very precise and sounds so low tech. We will employ more modern numerical methods.

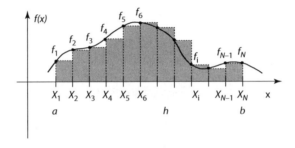

Figure 9.1 The rectangle method illustration. Shaded boxes show the approximations of the area under the curve. Here, $f_1 = f(x_1 = a), f_2 = f(x_2), f_3 = f(x_3), \ldots,$ $f_N = f(x_N = b)$.

Words of wisdom

Once you become proficient with computers, there is a tendency to do every problem numerically. Resist it! If you can find an analytical solution for a problem, do so. It is usually faster to calculate, and more importantly, will provide you with some physical intuition for the problem overall. Numerical solutions usually do not possess predictive power.

9.2 The Rectangle Method

The definition of the integral via Riemann's sum:

$$\int_a^b f(x)dx = \lim_{N \to \infty} \sum_{i=1}^{N-1} f(x_i)(x_{i+1} - x_i) \tag{9.1}$$

where:

$a \leq x_i \leq b$

N is the number of points

Riemann's definition gives us directions for the rectangle method: approximate the area under the curve by a set of boxes or rectangles (see Figure 9.1). For simplicity, we evaluate our function at N points equally separated by the distance h on the interval (a, b).

Rectangle method integral estimate

$$\int_a^b f(x)dx \approx \sum_{i=1}^{N-1} f(x_i)h \tag{9.2}$$

where

$$h = \frac{b-a}{N-1}, \quad x_i = a + (i-1)h, \quad x_1 = a \text{ and } x_N = b \tag{9.3}$$

The MATLAB implementation of this method is quite simple:

Listing 9.1 `integrate_in_1d.m` (available at `http://physics.wm.edu/programming_with_MATLAB_book/./ch_integration/code/integrate_in_1d.m`)

```
function integral1d = integrate_in_1d(f, a, b)
% integration with simple rectangle/box method
% int_a^b f(x) dx

N=100; % number of points in the sum
x=linspace(a,b,N);

s=0;
for xi=x(1:end-1) % we need to exclude x(end)=b
    s = s + f(xi);
end

%% now we calculate the integral
integral1d = s*(b-a)/(N-1);
```

To demonstrate how to use it, let's check $\int_0^1 x^2 dx = 1/3$.

```
f = @(x) x.^2;
integrate_in_1d(f,0,1)
ans =   0.32830
```

Well, 0.32830 is quite close to the expected value of 1/3. The small deviation from the exact result is due to the relatively small number of points; we used $N = 100$.

If you have previous experience in low-level languages (from the array functions implementation point of view) such as C or Java, this implementation is the first thing that will come to your mind. While it is correct, it does not employ MATLAB's ability to use matrices as arguments of a function. A better way, which avoids using a loop, is shown in Listing 9.2.

Listing 9.2 `integrate_in_1d_matlab_way.m` (available at `http://physics.wm.edu/programming_with_MATLAB_book/./ch_integration/code/integrate_in_1d_matlab_way.m`)

```
function integral1d = integrate_in_1d_matlab_way(f, a, b)
% integration with simple rectangle/box method
% int_a^b f(x) dx

N=100; % number of points in the sum
x=linspace(a,b,N);

% if function f can work with vector argument then we can
    do
```

```
integral1d = (b-a)/(N-1)*sum( f( x(1:end-1) ) ) ); % we
    exclude x(end)=b
```

Words of wisdom

In MATLAB, loops are generally slower compared with the equivalent evaluation of a function with a vector or matrix argument. Try to avoid loops that iterate evaluation over matrix elements.

9.2.1 Rectangle method algorithmic error

While the rectangle method is very simple to understand and implement, it has awful algorithmic error and slow convergence. Let's see why this is so. A closer look at Figure 9.1 reveals that a box often underestimates (e.g., the box between x_1 and x_2) or overestimates (e.g., the box between x_6 and x_7) the area. This is the algorithmic error, which is proportional to $f'h^2/2$ to the first order, that is, the area of a triangle between a box and the curve. Since we have $N-1$ intervals, this error accumulates in the worst-case scenario. So, we can say that the algorithmic error (E) is

Rectangle method algorithmic error estimate

$$E = \mathcal{O}\left((N-1)\frac{h^2}{2}f'\right) = \mathcal{O}\left(\frac{(b-a)^2}{2N}f'\right) \tag{9.4}$$

In the last term, we replaced $N-1$ with N under the assumption that N is large. The $\mathcal{O}$ symbol represents the *big O notation*, that is, there is an unknown proportionality coefficient.

9.3 Trapezoidal Method

The approximation of each interval with a trapezoid (see Figure 9.2) is an attempt to circumvent the weakness of the rectangle method. In other words, we do a linear approximation of the integrated function. Recalling the formula for the area of a trapezoid, we approximate the area of each interval as $h(f_{i+1} + f_i)/2$, and then we sum them all.

Trapezoidal method integral estimate

$$\int_a^b f(x)dx \approx h \times \left(\frac{1}{2}f_1 + f_2 + f_3 + \cdots + f_{N-2} + f_{N-1} + \frac{1}{2}f_N\right) \tag{9.5}$$

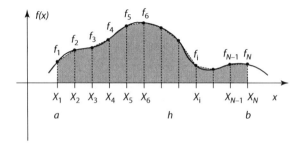

Figure 9.2   The illustration of the trapezoidal method. Shaded trapezoids show the approximations of the area under the curve.

The 1/2 coefficient disappears for inner points, since they are counted twice: once for the left and once for the right trapezoid area.

9.3.1   Trapezoidal method algorithmic error

To evaluate the algorithmic error, one should note that we are using a linear approximation for the function and ignoring the second-order term. Recalling the Taylor expansion, this means that to the first order, we are ignoring the contribution of the second derivative (f'') terms. With a bit of patience, it can be shown that

Trapezoidal method algorithmic error estimate

$$E = \mathcal{O}\left(\frac{(b-a)^3}{12N^2}f''\right)$$ (9.6)

Let's compare the integral estimate with rectangle (Equation 9.2) and trapezoidal (Equation 9.5) methods.

$$\int_a^b f(x)dx \approx h \times (f_2 + f_3 + \cdots + f_{N-2} + f_{N-1}) + h \times (f_1),$$ (9.7)

$$\int_a^b f(x)dx \approx h \times (f_2 + f_3 + \cdots + f_{N-2} + f_{N-1}) + h \times \frac{1}{2}(f_1 + f_N)$$ (9.8)

It is easy to see that they are almost identical, with the only difference in the second term of either $h \times (f_1)$ for the rectangle method or $h \times \frac{1}{2}(f_1 + f_N)$ for the trapezoidal one. While this might seem like a minor change in the underlying formula, it results in the algorithmic error decreasing as $1/N^2$ for the trapezoidal method, which is way better than the $1/N$ dependence for the rectangle method.

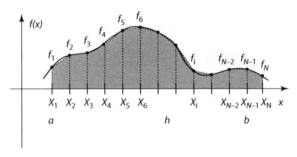

Figure 9.3   Simpson's method illustration.

9.4 Simpson's Method

The next logical step is to approximate the function with second-order curves, that is, parabolas (see Figure 9.3). This leads us to the Simpson method. Let's consider a triplet of consequent points (x_{i-1}, f_{i-1}), (x_i, f_i), and (x_{i+1}, f_{i+1}). The area under the parabola passing through all of these points is $h/3 \times (f_{i-1} + 4f_i + f_{i+1})$. Then, we sum the areas of all triplets and obtain

---

Simpson's method integral estimate

$$\int_a^b f(x)dx \approx h\frac{1}{3} \times (f_1 + 4f_2 + 2f_3 + 4f_4 + \cdots + 2f_{N-2} + 4f_{N-1} + f_N) \quad (9.9)$$

N must be in special form $N = 2k + 1$, that is, odd, and ≥ 3

Yet again, the first (f_1) and last (f_N) points are special, since they are counted only once, while the edge points $f_3, f_5, \ldots$ are counted twice as members of the left and right triplets.

9.4.1 Simpson's method algorithmic error

Since we are using more terms of the function's Taylor expansion, the convergence of the method is improving, and the error of the method decreases with N even faster.

Simpson method algorithmic error estimate

$$E = \mathcal{O}\left(\frac{(b-a)^5}{180N^4} f^{(4)}\right) \quad (9.10)$$

One might be surprised to see $f^{(4)}$ and N^4. This is because when we integrate a triplet area, we go by the h to the left and to the right of the central point, so the terms proportional to the $x^3 \times f^{(3)}$ are canceled out.

9.5   Generalized Formula for Integration

A careful reader may have already noticed that the integration formulas for these previous methods can be written in the same general form.

Generalized formula for numerical integration

$$\int_a^b f(x)\,dx \approx h \sum_{i=1}^{N} f(x_i)w_i \tag{9.11}$$

where w_i is the weight coefficient.

So, one has no excuse to use the rectangle and trapezoid method over Simpson's, since the calculational burden is exactly the same, but the calculation error for Simpson's method drops drastically as the number of points increases.

Strictly speaking, even Simpson's method is not so superior in comparison with others that use even higher-order polynomial approximation for the function. These higher-order methods would have exactly the same form as Equation 9.11. The difference will be in the weight coefficients. One can see a more detailed discussion of this issue in [9].

9.6   Monte Carlo Integration

9.6.1   Toy example: finding the area of a pond

Suppose we want to estimate the area of a pond. Naively, we might rush to grab a measuring tape. However, we could just walk around and throw stones in every possible direction, and draw an imaginary box around the pond to see how many stones landed inside (N_{inside}) the pond out of their total number (N_{total}). As illustrated in Figure 9.4, the ratio of these numbers should be proportional to the ratio of the pond area to the box area. So, we conclude that the estimated area of the pond (A_{pond})

$$A_{pond} = \frac{N_{inside}}{N_{total}} A_{box} \tag{9.12}$$

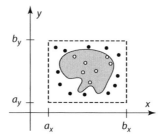

Figure 9.4   The pond area estimate via the Monte Carlo method.

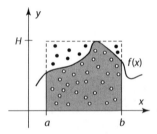

Figure 9.5 Estimate of the integral by counting points under the curve and in the surrounding box.

where:

$$A_{box} = (b_x - a_x)(b_y - a_y)$$

is the box area. The above estimate requires that the thrown stones are distributed *randomly* and *uniformly*.* It is also a good idea to keep the area of the surrounding box as tight as possible to increase the probability of hitting the pond. Imagine what would happen if the box were huge compared with the pond size: we would need a lot of stones (while we have only a finite amount of them) to hit the pond even once, and until then, the pond area estimate would be zero in accordance with Equation 9.12.

9.6.2 Naive Monte Carlo integration

Well, the calculation of the area under the curve is not much different from the measurement of the pond area as shown in Figure 9.5. So,

$$\int_{a_x}^{b_x} f(x)dx = \frac{N_{inside}}{N_{total}} A_{box}$$

9.6.3 Monte Carlo integration derived

The method described in the previous subsection is not optimal. Let's focus our attention on a narrow strip around the point x_b (see Figure 9.6). For this strip,

$$\frac{N_{inside}}{N_{total}} H \approx f(x_b) \tag{9.13}$$

So, there is no need to waste all of these resources if we can get the $f(x_b)$ right away. Thus, the improved estimate of the integral by the Monte Carlo method looks like

---

* This is not a trivial task. We will talk about this in Chapter 11. For now, we will use the rand function provided by MATLAB.

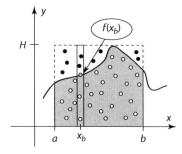

Figure 9.6   Explanation for the improved Monte Carlo method.

The Monte Carlo method integral estimate

Choose N uniformly and randomly distributed points x_i inside $[a, b]$

$$\int_a^b f(x)dx \approx \frac{b-a}{N} \sum_{i=1}^{N} f(x_i) \qquad (9.14)$$

9.6.4   The Monte Carlo method algorithmic error

A careful look at Equation 9.14 shows that $1/N \sum_{i=1}^{N} f(x_i) = \langle f \rangle$ is actually a statistical estimate of the function mean. Statistically, we are only sure of this estimate within the standard deviation of the mean estimate. This brings us to the following expression.

Monte Carlo method algorithmic error estimate

$$E = \mathcal{O}\left(\frac{b-a}{\sqrt{N}} \sqrt{\langle f^2 \rangle - \langle f \rangle^2}\right) \qquad (9.15)$$

where

$$\langle f \rangle = \frac{1}{N} \sum_{i=1}^{N} f(x_i)$$

$$\langle f^2 \rangle = \frac{1}{N} \sum_{i=1}^{N} f^2(x_i)$$

The rightmost term under the square root is the estimate of the function standard deviation $\sigma = \sqrt{\langle f^2 \rangle - \langle f \rangle^2}$.

Looking at Equation 9.15, you might conclude that you have wasted precious minutes of your life reading about a very mediocre method, which reduces its error proportionally to $1/\sqrt{N}$. This is worse even compared with the otherwise awful rectangle method.

Hold your horses. In the following section, we will see how the Monte Carlo method outshines all others for the case of multidimensional integration.

9.7 Multidimensional Integration

If someone asks us to calculate a multidimensional integral, we just need to apply our knowledge of how to deal with single dimension integrals. For example, in the two-dimensional case, we simply rearrange terms:

$$\int_{a_x}^{b_x} \int_{a_y}^{b_y} f(x,y)\, dx\, dy = \int_{a_x}^{b_x} dx \int_{a_y}^{b_y} dy\, f(x,y) \qquad (9.16)$$

The last single dimension integral is the function of only x:

$$\int_{a_y}^{b_y} dy\, f(x,y) = F(x) \qquad (9.17)$$

So, the two-dimensional integral boils down to the chain of two one-dimensional ones, which we are already fit to process:

$$\int_{a_x}^{b_x} \int_{a_y}^{b_y} f(x,y)\, dx\, dy = \int_{a_x}^{b_x} dx\, F(x) \qquad (9.18)$$

9.7.1 Minimal example for integration in two dimensions

Listing 9.3 shows how to do two-dimensional integrals by chaining the one-dimensional ones; note that it piggybacks on the single-integral in Listing 9.1 (but feel free to use any other method).

Listing 9.3 integrate_in_2d.m (available at http://physics.wm.edu/programming_with_MATLAB_book/./ch_integration/code/integrate_in_2d.m)

```
function integral2d=integrate_in_2d(f, xrange, yrange)
% Integrates function f in 2D space
% f is handle to function of x, y i.e. f(x,y) should be valid
% xrange is a vector containing lower and upper limits of integration
%    along the first dimension.
%    xrange = [x_lower x_upper]
% yrange is similar but for the second dimension
```

```
% We will define (Integral f(x,y) dy) as Fx(x)
Fx = @(x) integrate_in_1d( @(y) f(x,y), yrange(1), yrange(2) );
%    ^^^^^ we fix 'x',        ^^^^^^^^^^^here we reuse this already fixed
   x
%                                       so it reads as Fy(y)
% This is quite cumbersome.
% It is probably impossible to do a general D-dimensional case.
% Notice that matlab folks implemented integral, integral2, integral3
% but they did not do any for higher than 3 dimensions.

integral2d = integrate_in_1d(Fx, xrange(1), xrange(2) );
end
```

Let's calculate

$$\int_0^2 dx \int_0^1 (2x^2 + y^2) \, dy \tag{9.19}$$

```
f = @(x,y) 2*x.^2 + y.^2;
integrate_in_2d( f, [0,2], [0,1] )
ans =   5.9094
```

It is easy to see that the exact answer is 6. The observed deviation from the
analytical result is due to the small number of points used in the calculation.

9.8 Multidimensional Integration with Monte Carlo

The "chain" method in the previous subsection can be expanded to any number
of dimensions. Can we rest now? Not so fast. Note that if we would like to split
the integration region by N points in each of the D dimensions, then the number
of evaluations, and thus the calculation time, grows $\sim N^D$. This renders the rectangle, trapezoidal, Simpson's, and similar methods useless for high-dimension
integrals.

The Monte Carlo method is a notable exception; it looks very simple even for
a multidimensional case, it maintains the same $\sim N$ evaluation time, and its error
is still $\sim 1/\sqrt{N}$.

A three-dimensional case, for example, would look like this:

$$\int_{a_x}^{b_x} dx \int_{a_y}^{b_y} dy \int_{a_z}^{b_z} dz \, f(x,y,z) \approx \frac{(b_x - a_x)(b_y - a_y)(b_z - a_z)}{N} \sum_{i=1}^{N} f(x_i, y_i, z_i) \tag{9.20}$$

and the general form is shown in the following box:

Monte Carlo method in D-space

$$\int_{V_D} dV_D \, f(\vec{x}) = \int_{V_D} dx_1 dx_2 dx_3 ... dx_D \, f(\vec{x}) \approx \frac{V_D}{N} \sum_{i=1}^{N} f(\vec{x}_i) \qquad (9.21)$$

where:

V_D is the D-dimensional volume

$\vec{x}_i$ are randomly and uniformly distributed points in the volume V_D

9.8.1 Monte Carlo method demonstration

To see how elegant and simple the implementation of the Monte Carlo method can be, we will evaluate the integral in Equation 9.19.

```
f = @(x,y) 2*x.^2 + y.^2;
bx=2; ax=0;
by=1; ay=0;
% first we prepare x and y components of random points
% rand provides uniformly distributed points in the (0,1)
    interval
N=1000; % Number of random points
x=ax+(bx-ax)*rand(1,N);   % 1 row, N columns
y=ay+(by-ay)*rand(1,N);   % 1 row, N columns

% finally integral evaluation
integral2d = (bx-ax)*(by-ay)/N * sum( f(x,y) )
integral2d =   6.1706
```

We used only 1000 points, and the result is quite close to the analytical value of 6.

9.9 Numerical Integration Gotchas

9.9.1 Using a very large number of points

Since the algorithmic error of numerical integration methods drops with an increase of N, it is very tempting to increase it. But we must remember about round-off errors and resist this temptation. So, h should not be too small, or equivalently, N should not be too big. N definitely should not be infinite as the Riemann's sum in Equation 9.1 prescribes, since our lifetime is finite.

9.9.2 Using too few points

There is a danger of under-sampling if the integrated function changes very quickly and we put points very sparsely (see e.g., Figure 9.7). We should first plot

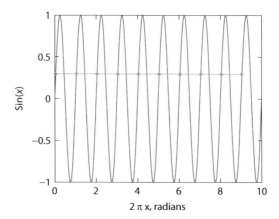

Figure 9.7 The example of badly chosen points (○) for integration of the $\sin(x)$ function (−). The samples give the impression that the function is a horizontal line.

the function (if it is not too computationally taxing) and then choose an appropriate number of points: there should be at least two and ideally more points for every valley and hill of the function.* To see if you hit a sweet spot, try to double the amount of points and see whether the estimate of the integral stops changing drastically. This is the foundation for the so-called *adaptive integration* method, where an algorithm automatically decides on the required number of points.

9.10   MATLAB Functions for Integration

There follows a short summary of MATLAB's built-in functions for integration:
    One-dimensional integration

- `integral`.
- `trapz` employs modified trapezoidal method.
- `quad` employs modified Simpson's method.

Here is how to use quad to calculate $\int_0^2 3x^3 dx$:

```
>> f = @(x) 3*x.^2;
>> quad(f, 0, 2)
ans =
    8.0000
```

As expected, the answer is 8. Note that `quad` expects your function to be vector friendly.
    Multidimensional integration

- `integral2` two-dimensional case
- `integral3` three-dimensional case

---

* We will discuss it in detail in the section devoted to the discrete Fourier transform (Chapter 15).

Let's calculate the integral $\int_0^1 \int_0^1 (x^2 + y^2)dxdy$, which equals 2/3.

```
>> f = @(x,y) x.^2 + y.^2;
>> integral2( f, 0, 1, 0, 1)
ans =
    0.6667
```

There are many other built-ins. See MATLAB's numerical integration documentation to learn more.

MATLAB's implementations are more powerful than those that we discussed, but deep inside, they use similar methods.

9.11 Self-Study

General comments:

- Do not forget to run some test cases.

- MATLAB has built-in numerical integration methods, such as quad. You might check the validity of your implementations with answers produced by this MATLAB built-in function. quad **requires your function to be able to work with an array of** x **points**, otherwise it will fail.

  - Of course, it is always better to compare with the exact analytically calculated value.

Problem 9.1
Implement the trapezoidal numerical integration method. Call your function trapezInt(f,a,b,N), where a and b are left and right limits of integration, N is the number of points, and f is the handle to the function.

Problem 9.2
Implement the Simpson numerical integration method. Call your function simpsonInt(f,a,b,N). Remember about special form of $N = 2k + 1$.

Problem 9.3
Implement the Monte Carlo numerical integration method. Call your function montecarloInt(f,a,b,N).

Problem 9.4
For your tests, calculate

$$\int_0^{10} [\exp(-x) + (x/1000)^3]dx$$

Plot the integral absolute error from its true value for the above methods (include rectangular method as well) vs. different number of points N. Try to do it from small $N = 3$ to $N = 10^6$. Use loglog plotting function for better representation (make sure that you have enough points in all areas of the plot). Can you relate the trends to eqs. (9.4), (9.6), (9.9), and (9.15)? Why does the error start to grow with a larger N? Does it grow for all methods? Why?

Problem 9.5

Calculate

$$\int_0^{\pi/2} \sin(401x)\,dx$$

Compare your result with the exact answer 1/401. Provide a discussion about the required number of points to calculate this integral.

Problem 9.6

Calculate

$$\int_{-1}^1 dx \int_0^1 dy\,(x^4 + y^2 + x\sqrt{y})$$

Compare results of the Monte Carlo numerical integration method and MATLAB's built-in integral2.

Problem 9.7

Implement a method to find the volume of the N-dimensional sphere for the arbitrary dimension N and the sphere radius R:

$$V(N,R) = \iiint_{x_1^2 + x_2^2 + x_3^2 + \cdots + x_N^2 \leq R^2} dx_1 dx_2 dx_3 \cdots dx_N$$

Calculate the volume of the sphere with $R = 1$ and $N = 20$.

CHAPTER 10

Data Interpolation

In this chapter, we cover several of the most common methods for interpolation. Our assumption is that we have a data set of $\{x\}$ and $\{y\}$ data points. Our job is to provide some algorithm that will find the interpolated yi position for any point xi that is located in between the known data point positions along the x-axis.

There is rarely enough data, but it often takes a lot of time, money, and effort to get even a single data point. However, we would like to have some representation of the measured system in the voids between data points. The process of artificially filling such voids is called *data interpolation*. Often, people confuse fitting (see Chapter 6) and interpolation. These are two fundamentally different operations. The fitting tells us what the data in the voids should look like, since it assumes the knowledge of the underlying model, while the interpolation tells us what it might look like. Also, it is necessary that interpolated lines pass through the data points by design, while this is not necessary for fitting. Whenever possible, fitting should be preferred over interpolation. We should reserve interpolation for the processes for which we do not know the underlying equations or if fitting takes too much time.

10.1 The Nearest Neighbor Interpolation

The name says it all. For each point to be interpolated, xi, we find the nearest neighbor along the x-axis in the data set and use its y value. We implement the *nearest neighbor interpolation* with MATLAB in the annotated code in Listing 10.1.

Listing 10.1 `interp_nearest.m` (available at `http://physics.wm.edu/programming_with_MATLAB_book/./ch_interpolation/code/interp_nearest.m`)

```
function yi= interp_nearest(xdata,ydata,xi)
%% Interpolates by the nearest neighbor method
% xdata, ydata - known data points
% xi - points at which we want the interpolated values yi
% WARNING: no checks is done that xi values are inside xdata range

%% It is crucial to have yi preinitialized !
% It significantly speeds up the algorithm,
% since computer does not have to reallocate memory for  new data points.
% Try to comment the following line and compare the execution time
% for a large length of xi
yi=0.*xi; % A simple shortcut to initialize return vector with zeros.
          % This also takes care of deciding the yi vector type (row or column).

%% Finally, we interpolate.
N=length(xi);
for i=1:N % we will go through all points to be interpolated
    distance=abs(xdata-xi(i));
    % MATLAB's min function returns not only the minimum but its index too
```

```
      [distance_min, index] = min(distance);
      % there is a chance that 2 points of xdata have the same distance to the xi
      % so we will take the 1st one
      yi(i)=ydata(index(1));
  end
end
```

In this code, we used the feature of MATLAB's min function, which can return not only the minimum of an array but also the index or location of it in the array. Let's use the data points in Listing 10.2 for this and the following examples.

Listing 10.2 data_for_interpolation.m (available at http://physics.wm. edu/programming_with_MATLAB_book/./ch_interpolation/code/ data_for_interpolation.m)

```
xdata=[-1,   0,      3, 1,      5,  6, 10,   8];
ydata=[-2,   0.5,   4, 1.5,   8,  6, 2,    3];
```

The interpolation points are obtained with

```
Np=300; % number of the interpolated points
xi=linspace(min(xdata), max(xdata),Np);
yi=interp1(x,y,xi,'nearest');
```

The plot of the data and its interpolation by the nearest neighbor method is shown in Figure 10.1.

10.2   Linear Interpolation

The curve interpolated with the nearest neighbor method has many discontinuities, which is not visually appealing. The *linear interpolation* method mitigates this issue.

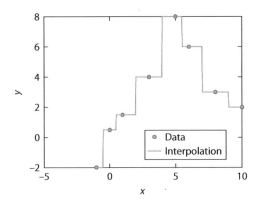

Figure 10.1   Data and its interpolation with the nearest neighbor method.

We will split our data set with N points in to $N-1$ intervals and interpolate the values in the given interval as a line passing through the border points (x_i, y_i) and (x_{i+1}, y_{i+1}). This is implemented with the code shown in Listing 10.3.

Listing 10.3 `interp_linear.m` (available at `http://physics.wm.edu/` `programming_with_MATLAB_book/./ch_interpolation/code/` `interp_linear.m`)

```
function yi= interp_linear(xdata,ydata,xi)
%% Interpolates with the linear interpolation method
% xdata, ydata - known data points
% xi - points at which we want interpolated values yi
% WARNING: no checks is done that xi values are inside xdata range

% First, we need to sort our input vectors along the x coordinate.
% We need the monotonous growth of x

% MATLAB's sort function has an extra return value: the index.
% The list of indexes is such that x=xdata(index), where x is sorted
[x,index]=sort(xdata);
% We reuse this index to sort 'y' vector the same way as 'x'
y=ydata(index);

%% Second we want to calculate parameters of the connecting lines.
% For N points we will have N-1 intervals with connecting lines.
% Each of them will have 2 parameters slope and offset,
% so we need the parameters matrix with Nintervals x 2 values
Nintervals=length(xdata)-1;
p=zeros(Nintervals,2);
% p(i, 1) is the slope  for the interval between x(i) and x(i+1)
% p(i, 2) is the offset for the interval between x(i) and x(i+1)
% so y = offset*x+slope = p1*x+p2 at this interval
for i=1:Nintervals
    slope = ( y(i+1)-y(i) ) / ( x(i+1)-x(i) );  % slope
    offset = y(i)-slope*x(i);  % offset at x=0
    p(i,1)=slope;
    p(i,2)=offset;
end

%% It is crucial to have yi preinitialized !
% It significantly speeds up the algorithm,
% since computer does not have to reallocate memory for  new data
   points.
% Try to comment the following line and compare the execution time
% for a large length of xi
yi=0.*xi; % A simple shortcut to initialize return vector with zeros.
          % This also takes care of deciding the yi vector type (row or
              column).

%% Finally, we interpolate.
N=length(xi);
for i=1:N % we will go through all points to be interpolated
    % Let's find nearest left neighbor for xi.
    % Such neighbor must have the smallest positive displacement
    displacement=(xi(i)-x);
```

```
    [displ_min_positive, interval_index] = min( displacement(
        displacement >= 0 ));
    if (interval_index > Nintervals )
        % We will reuse the last interval parameters.
        % Since xi must be within the xdata range, this the case
        % when xi=max(xdata).
        interval_index = Nintervals;
    end
    % The index tells which  interval to use for the linear
        approximation.
    % The line is the polynomial of the degree 1.
    % We will use the MATLAB's 'polyval' function
    % to evaluate value of the polynomial of the degree n
    % at point x: y=p_1*x^n+p_2*x^(n-1)+ p_n*x +p_{n+1}
    % yi(i)= p(interval_index,1)  * xi(i) +p(interval_index,2);
    poly_coef=p(interval_index,:);
    yi(i)=polyval(poly_coef,xi(i));
end
end
```

The results of the linear interpolation are shown in Figure 10.2. By the way, MATLAB uses the linear interpolation in the plot command when you ask it to join points with lines.

In the code in Listing 10.3, we used MATLAB's polyval function, which calculates the value of the linear polynomial. This is an overkill for the simple line, but we will see its use in the following section in a more elaborate situation.

10.3 Polynomial Interpolation

The linear interpolation looks nicer than the nearest neighbor one, but the interpolated curve is still not smooth. The next natural step is to use a higher-degree polynomial, which is definitely smooth, that is, it does not have discontinuities of the first derivative.

We will do it with two of MATLAB's built-in functions: polyfit and polyval. The polyfit finds the coefficient of the $N-1$ degree polynomial passing through

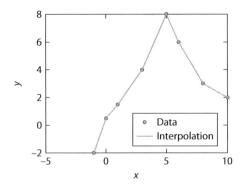

Figure 10.2   Data and its interpolation with the linear interpolation method.

N data points.* The `polyval` function finds the value of the polynomial with its coefficient provided in the p array according to the following formula:

$$P_N(x) = p_1 x^N + p_2 x^{N-1} + \cdots + p_N x + p_{N+1} \tag{10.1}$$

The resulting code for the *polynomial interpolation* is

```
% calculate polynomial coefficients
p=polyfit(xdata, ydata, (length(xdata)-1) );
%  interpolate
yi=polyval(p,xi);
```

The results of the polynomial interpolation are shown in Figure 10.3. With this method, the interpolated values tend to oscillate, especially for a polynomial of a high degree. We can see precursors of this oscillation at the right end of the interpolation for x values between 8 and 10. The interpolated line swings quite high from an imaginary smooth curve through the points of a linearly interpolated curve (see Figure 10.2). While this is not strictly a bad thing, since we do not know what the data look like in between the data points anyway, natural processes rarely show such swings.

The real problem is that the polynomial interpolation is highly sensitive to the addition of new data points. See how drastically the addition of just one more point $(2, 1)$ changes the interpolated curve in Figure 10.4. It is clear that the oscillatory behavior is enhanced. Also, a careful comparison of Figure 10.3 and Figure 10.4 reveals that the interpolated line changed everywhere. This is clearly an unwelcome feature.

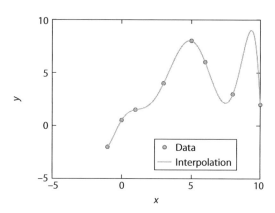

Figure 10.3   Data and their interpolation with the polynomial interpolation method.

---

* We can always find a polynomial of $N-1$ degree passing through N data points, since we have enough data to form N equations for N unknown polynomial coefficients. We can do this with methods described in Chapter 5.

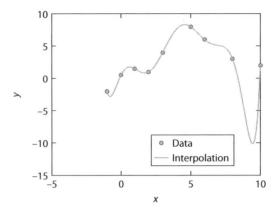

Figure 10.4   Data and their interpolation with the polynomial interpolation method with one more additional point at $(2, 1)$.

Words of wisdom

Stay away from high-order polynomial interpolation. It brings nothing but trouble.

10.4   Criteria for a Good Interpolation Routine

A good interpolation routine should be robust to the addition of new data points, that is, new data points should modify the interpolation curve at non-adjacent intervals as little as possible. Ideally, there should be no change at all. Otherwise, we would have to recalculate the interpolation curve and related coefficient every time after the addition of a point. The nearest neighbor and linear interpolation methods satisfy this criterion.

10.5   Cubic Spline Interpolation

One more method that produces a nice, smooth interpolation curve and is relatively immune to the addition of new data points is called the *cubic spline interpolation* method. In the old days, this method used to be implemented literally with hardware. Whenever a smooth curve was required, people would use pegs or knots to fix the location of an elastic ruler (or spline) that passed through knots. By the "magic" of the elastic material behavior, the connection was continuous and smooth.

Nowadays, we mimic the hardware counterpart by setting the underlying equations. We use the third-degree polynomial for each ith interval between the two adjacent data points as the interpolation curve.

$$f_i(x) = p_{1_i}x^3 + p_{2_i}x^2 + p_{3_i}x + p_{4_i}, x \in [x_i, x_{i+1}] \tag{10.2}$$

We require that the interpolating polynomial passes through the interval's border data points:

$$f_i(x_i) = y_i \tag{10.3}$$

$$f_i(x_{i+1}) = y_{i+1} \tag{10.4}$$

Eqs. (10.3) and (10.4) are not sufficient to constrain the four polynomial coefficients. So, as additional constraints, we request $f_i(x)$ to have continuous first-order derivative at the interval's borders to minimize the bending of the ruler at knots:

$$f'_{i-1}(x_i) = f'_i(x_i) \tag{10.5}$$

$$f'_i(x_{i+1}) = f'_{i+1}(x_{i+1}) \tag{10.6}$$

These four equations are sufficient to find four unknown coefficients of the polynomial at every interval, except the leftmost and rightmost ones, since they are missing a neighboring interval. For the end points, we can choose any constraint we like for the first or second derivative. The physical ruler was free to move at the end knots so that it would look like a straight line (i.e., no bending) at the end points. Thus, the border conditions for the *natural spline* are to set the second derivatives at the ends to zero.

$$f''_1(x_1) = 0 \tag{10.7}$$

$$f''_{N-1}(x_N) = 0 \tag{10.8}$$

Then, we offload the task of solving four equations at every interval to a computer and find the resulting interpolation curve. The result is shown in Figure 10.5. As you can see, the spline interpolation is smooth and continuous and "naturally" follows the data. The algorithm for the cubic interpolation method is not shown here; it is not very complex, though it involves quite a lot of bookkeeping. The author's opinion is that we can spend our time on something more exciting and use MATLAB's built-in implementation.

10.6   MATLAB Built-In Interpolation Methods

All of the interpolation methods in the previous section are implemented in MATLAB's built-in:

```
interp1(xdata, ydata, xi, method)
```

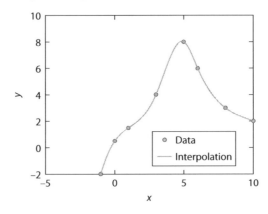

Figure 10.5 Data and their interpolation with the natural cubic spline interpolation method with one more additional point at $(2, 1)$.

Here, the method could be

- linear for the linear interpolation (default)
- nearest for the nearest neighbor interpolation
- spline for the cubic spline interpolation

See other methods and options in the help file.

10.7 Extrapolation

Extrapolation is the process of filling voids outside of a measured region. The MATLAB built-ins allow us to send an additional argument to obtain the extrapolated points.

However, the author is convinced that **extrapolation must be avoided** at any cost.* The only exception is when we have a model of the underlying process.

10.8 Unconventional Use of Interpolation

10.8.1 Finding the location of the data crossing $y = 0$

Suppose we have a bunch of data points, and we would like to estimate where the underlying process crosses the $y = 0$ line. We can mimic the data generating process with any interpolation of our choice, for example, the cubic spline.

```
>> fi = @(xi) interp1(xdata, ydata, xi, 'linear')
```

---

* Before satellites were available to monitor weather, meteorologists spent a great deal of money and effort to put manned weather stations close to the north and south poles to avoid extrapolation and produce reliable weather forecasts.

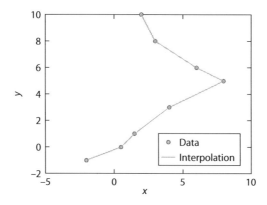

Figure 10.6   The flipped data and the flipped linear interpolated curve.

Then, we can use a variety of the root finding methods described in Chapter 8 to find the answer. For example,

```
>> xguess=0;
>> xi0= fsolve( fi, xguess)
xi0 = -0.2000
```

However, this is a highly inefficient method, since it takes many evaluations of the interpolation function during the iterative root finding process. Therefore, we must resist the temptation to use it.

A better way is to mentally flip xdata and ydata, then find the interpolated value at the yi = 0 point. This can all be done in just one line:

```
xi0 = interp1(ydata, xdata, 0, 'linear')
```

There is one more caveat: the flipped curve must have a single value for each new x, which is not the case for our example data, as can be seen in Figure 10.6. So, we must constrain the data set in the vicinity of the "root". In our examples, we should use the non-flipped points with $-3 \leq x_{data} \leq 5$ (see Figure 10.2). So, our root estimating code is just a bit more complex.

```
>> indx = ( -3 <= xdata ) & ( xdata <=5 ); % which x
   indices satisfy the constrains
>> x_constrained = xdata(indx);
>> y_constrained = ydata(indx);
>> xi0 = interp1(y_constrained,x_constrained,0,'linear')
xi0 = -0.2000
```

If you use different interpolation techniques, the roots found by the first and second method may differ, since the constrained data might have different bending in the interpolation. We should not be too worried about it, since this is only an estimate. In any case we have no clue what the data look like in the voids, so any estimate is a good estimate, although the linear spline inter-polation will produce the same result, that is, it is robust to the root estimate method.

10.9   Self-Study

Problem 10.1
Using MATLAB's interp1 with spline method, find where the interpolation line crosses the $y = 0$ level. The interpolation is done over the following points [(x,y) notation]: $(2, 10)$, $(3, 8)$, $(4, 4)$, $(5, 1)$, $(6, -2)$.
     Would it be wise to search crossing with $x = 0$ line using these data? Why?

Problem 10.2
Plot results of the cubic spline interpolation for original data (see Listing 10.2) and the same data with the addition of one more point at (2,1). Is this method robust to addition of new data? Would you recommend it over the polynomial interpolation method?

Part III

Going Deeper and Expanding the Scientist's Toolbox

CHAPTER 11

Random Number Generators and Random Processes

This chapter provides an introduction to random number generators, methods to evaluate the quality of the generated "randomness," and MATLAB built-in commands to generate various random number distributions.

If we look around, we notice that many processes are nondeterministic, that is, we are not certain of their outcome. We are not 100% certain whether rain will happen tomorrow; the banks do not have certainty about their loan return; sometimes, we do not know whether our car will start or not. Our civilization makes the best effort to make life predictable, that is, we are quite certain that a roof will not leak tomorrow and a grocery store will have fruit for sale. Still, we cannot exclude the element of uncertainty from our calculation. To model such uncertain or random processes, we use random number generators (RNGs).

11.1 Statistics and Probability Introduction

11.1.1 Discrete event probability

Before we begin our discussion of RNGs, we need to set certain definitions. Suppose we record observations of a certain process that generates multiple discrete outcomes or events. Think, for example, about throwing a six-sided die that can produce numbers (outcomes) from 1 to 6. The mathematical and physical definition of the probability (p) of a discrete event x is given by

Probability of event ("x")

$$p_x = \lim_{N_{total} \to \infty} \frac{N_x}{N_{total}} \tag{11.1}$$

where:
  N_x is the number of registered event x
  N_{total} is the total number of all events

We would have to throw our die many (ideally infinite) times to see what is the probability of getting a certain number. This is very time consuming, so most of the time with a finite number of trials, we get only the *estimate* of the probability.[*]

---

[*] In some situations, we cannot assign a probability to an outcome at all, at least in the sense of Equation 11.1. For example, we cannot say what is the probability of finding life on a planet of Alpha Centauri; we have not yet done any measurements, that is, our N_{total} is 0. There are other ways to do it via conditional probabilities, but this is outside the scope of this chapter.

Sometimes, we can assign such probabilities if we know something about the process. For example, if we assume that our six-sided die is symmetric, there is no reason for one number to be more probable than any other; thus, the probability of any outcome is 1/6 for this die.

11.1.2 Probability density function

While in real life all events are discrete, mathematically it is convenient to work with events that are continuous, that is, no matter how small a spacing we choose, there is always an event possible next to the other one within this spacing. Think, for example, about the probability of pulling a random real number from the interval from 0 to 1. This interval (as well as any other non-zero interval) has an infinite amount of real numbers, and thus, the probability of pulling any particular number is zero.

In this situation, we should talk about the probability density of an event x, which is calculated by the following method: we split our interval into m equidistant bins, run our random process many times, and calculate

---

The probability density estimate of an event x

$$p(x) = \lim_{N_{total} \to \infty} \frac{N_{x_b}}{N_{total}}$$
(11.2)

where:
N_{x_b} is the number of events that land in the same bin as x
N_{total} is the total number of all events

As you can see from this definition, if we make number of bins (m) infinite and sum over all bins,

$$\sum_{i=1}^{m} N_i / N_{totol} = \int p(x)dx = 1$$
(11.3)

As in the case of probabilities of discrete events, sometimes we can assign the probability density distribution a priori.

11.2 Uniform Random Distribution

The uniform distribution is a very useful probability distribution. You can see its extensive use in Chapter 12 and Section 9.6. As the name suggests, the density function of this distribution is uniform, that is, it is the same everywhere. This means that the probability of pulling a number is the same in a given interval. For convenience, the default interval is set from 0 to 1. If you need a single number in this interval, just execute MATLAB's built-in rand function.

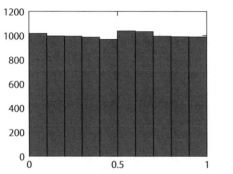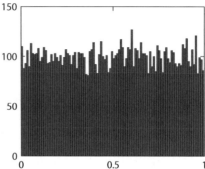

Figure 11.1   Histogram of $N = 10,000$ randomly and uniformly distributed events binned into $m = 10$ (left) and $m = 100$ (right) bins.

```
>> rand()
ans = 0.8147
```

Your result will be different, since we are dealing with random numbers.
    Let's check the uniformity of the MATLAB generator.

```
r=rand(1,N);
hist(r,m);
```

The first command generates N random numbers; the second splits our interval into m bins, counts how many times a random number gets into the given bin, and plots the histogram (thus the name hist), that is, number of events for every bin. You can see the results shown in Figure 11.1. It is clear that we hit a given bin roughly the same number of times, that is, the distribution is uniform. When the number of bins is relatively large in comparison to the number of events, we start to see bin-to-bin counts variation (as in the right panel of Figure 11.1). If we increase the number of random events, this bin-to-bin difference will reduce.

11.3   Random Number Generators and Computers

The word *random* means that we cannot predict the outcome based on previous information. How can a computer, which is very accurate, precise, and **deterministic**, generate random numbers? **It cannot!**

    The best we can do is to generate a sequence of *pseudo* random numbers. By "pseudo," we mean that starting from the same initial conditions, the computer will generate exactly the same sequence of numbers (*very handy for debugging*). But otherwise, the sequence will look like random numbers and will have the **statistical** properties of random numbers. In other words, if we do not know the RNG algorithm, we cannot predict the next generated number based on a known sequence of already produced numbers, while the numbers obey the required probability distribution.

11.3.1   Linear congruential generator

A very simple algorithm to generate uniformly distributed integer numbers in the 0 to $m-1$ interval is the *linear congruential generator* (LCG). It uses the following recursive formula:

LCG recursive formula

$$r_{i+1} = (a \times r_i + c) \bmod m \qquad\qquad (11.4)$$

where:
   m is the integer modulus
   a is the multiplier, $0 < a < m$
   c is the increment, $0 \le c < m$
   r_1 is the seed value, $0 \le r_1 < m$
mod is the modulus after division by m operation

All pseudo random generators have a period (see Section 11.3.2), and this one is no exception. Once r_i repeats one of the previous values, the sequence will repeat as well.

This LCG can have at most a period of m distinct numbers, since this is how many distinct outcomes the mod m operation has. A bad choice of a, c, m, r_1 will lead to an even shorter period.

Example

The LCG with these parameters $m = 10, a = 2, c = 1, r_1 = 1$ generates only 4 out of 10 possible distinct numbers: $r = [1, 3, 7, 5]$, and then the LCG repeats itself.

We show a possible realization of the LCG algorithm with MATLAB in Listing 11.1.

Listing 11.1 lcgrand.m (available at http://physics.wm.edu/ programming_with_MATLAB_book/./ch_random_numbers_generators/code/ lcgrand.m)

```
function r=lcgrand(Nrows,Ncols, a,c,m, seed)
% Linear Congruential Generator - pseudo random number
    generator

        r=zeros(Nrows, Ncols);
        r(1)=seed; % this equivalent to r(1,1)=seed;

        for i=2:Nrows*Ncols;
                r(i)= mod( (a*r(i-1)+c), m);
                % notice r(i)   and r(i-1)
```

```
                    % there is a way to address
                        multidimensional array
                    % with only one index
            end
            r=r/(m-1); %normalization to [0,1] interval
end
```

The LCG is fast and simple, but it is a very bad random numbers genera-
tor.[*] Sadly, quite a lot of numerical libraries still use it for historical reasons, so be
aware. Luckily for us, MATLAB uses a different algorithm by default.

11.3.2   Random number generator period

Even the best pseudo random generators cannot have a period larger than 2^B,
where B is the number of all available memory storage bits. This can be shown
by the following consideration. Suppose we had a particular bit combination, and
then we run an RNG and get a new random number. This must somehow modify the computer memory state. Since the memory bit can be only in an *on* or *off*
state, that is, there are only two states available, the total number of the different
memory states is 2^B. Sooner or later, the computer will go over all available memory states as the result of the RNG calculations, and the computer will have the
same state as it already had. At this point, the RNG will repeat the sequence of
generated numbers.

A typical period of an RNG is much smaller than 2^B, since it is unpractical to
use all available storage only for RNG needs. After all, we still need to do something useful beyond random number generations, that is, some other calculations
that will require memory too.

While the RNG period can be huge, it is not infinite. For example, MATLAB's
default RNG has the period $2^{19937} - 1$.

Why is this so important? Recall that the Monte Carlo integration method error
is $\sim 1/\sqrt{N}$ (see Section 9.6.4). This holds true only when N is less than the period
of the RNG used (T). For $N > T$, the Monte Carlo method cannot give uncertainty
better than $\sim 1/\sqrt{T}$, since it would sample the same T random numbers over and
over. To see this, suppose that someone wants to know an average public opinion.
He should choose many random people to ask their opinion. If he starts asking the
same two people over and over, it does not improve the estimate of public opinion.
Similarly, the Monte Carlo method will not improve for $N > T$: it will keep doing
calculations, but there will be no improvement in the result's precision.

11.4   How to Check a Random Generator

As we discussed in the previous section, computers alone are unable to generate
truly random numbers without additional hardware that uses the randomness of

---

[*] Do not use the LCG whenever your money or reputation is at stake!

some natural process (e.g., radioactive decay or quantum noise in the precision measurements). So, we should be concerned only with the RNG's statistical properties. If a given RNG generates a pseudo random sequence that has properties of random numbers, then we should be happy to use it.

The National Institute of Standards and Technology (NIST) provides software and several guidelines to check RNGs [2], though it is not easy, and probably impossible, to check all required random number properties.

11.4.1 Simple RNG test with Monte Carlo integration

If only the statistical properties of the RNG are important, we can test them by checking that the integral deviation calculated with the Monte Carlo algorithm drops by $1/\sqrt{N}$ from its true value.

In the code in Listing 11.2, we test properties of our LCG with the following coefficients: $m = 100$, $a = 2$, $c = 1$, and $r_1 = 1$. For our purposes, we can calculate the numerical integral estimate of any non-constant function; here, we calculate

$$\int_0^1 \sqrt{1-x^2}\, dx = \pi/4 \tag{11.5}$$

with the Monte Carlo algorithm and see whether the integration error drops by $1/\sqrt{N}$. For comparison, we use the MATLAB built-in algorithm (rand) as well. The code of this test is shown in Listing 11.2.

Listing 11.2 check_lcgrand.m (available at http://physics.wm.edu/ programming_with_MATLAB_book/./ch_random_numbers_generators/code/ check_lcgrand.m)

```
% We will calculate the deviation of the Monte Carlo
    numerical
% integration algorithm from the true integral value
% of the function below
f=@(x) sqrt(1-x.^2);
% integral of above on 0 to 1 interval is pi/4
% since we have shape of a quoter of the circle
trueInteralValue = pi/4;

Np=100; % number of trials
N=floor(logspace(1,6,Np)); % number of random points in
    each trial

% initialization of error arrays
erand=zeros(1,Np);
elcg =zeros(1,Np);

a = 2; c=1; m=100; seed=1;
```

```
for i=1:Np
    % calculate integration error
        erand(i)=abs( sum( f(rand(1,N(i))) )/N(i) -
            trueInteralValue);
        elcg(i) =abs( sum( f(lcgrand(1,N(i),a,c,m,seed)) )
            /N(i) - trueInteralValue);
end

loglog(N,erand,'o', N, elcg, '+');
set(gca,'fontsize',20);
legend('rand', 'lcg');
xlim([N(1),N(end)]);
xlabel('Number of requested random points');
ylabel('Integration error');
```

To run the test, we execute

```
check_lcgrand
```

The results of the comparison of the two RNG methods are shown in Figure 11.2. We can see that the integration errors keep dropping as we increase the number of points used by the Monte Carlo integration when we use MATLAB's rand RNG. Eyeballing the dependence of the error on N, we can see that the errors drop roughly by an order of magnitude as N increases by two orders of magnitude. This is typical for $1/\sqrt{N}$ behavior. Therefore, MATLAB's generator passes our check. The results with our LCG are quite different: the errors stop decreasing once we reach N around 100; by the time we reach $N \approx 1000$, the errors maintain an almost constant level. The position of the "elbow" for the LCG data roughly

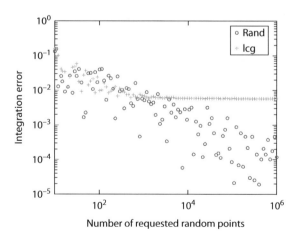

Figure 11.2   The comparison of Monte Carlo integration done with MATLAB's good built-in RNG (circles) and our LCG (crosses) with coefficients $m = 100$, $a = 2$, $c = 1$, and $r_1 = 1$.

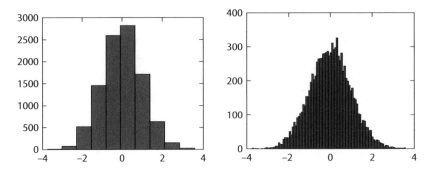

Figure 11.3   Histogram of 10,000 randomly and normally distributed events binned into 10 (left) and 100 (right) bins.

(within an order of magnitude) coincides with the period of the LCG. Recall that the period cannot be larger than 100 in this case, since $m = 100$.

11.5   MATLAB's Built-In RNGs

In this chapter, we focused our attention on the uniform pseudo RNG, which is implemented in MATLAB as the rand function. It generally takes two arguments: the number of rows (Nrows) and columns (Ncols) in the requested matrix of random numbers. The typical way to call it is as rand(Nrows, Ncols). For example,

```
>> rand(2, 3)
ans =
   0.1270      0.6324      0.2785
   0.9134      0.0975      0.5469
```

MATLAB can also produce random numbers distributed according to the standard normal distribution. Use randn for this. Compare the histograms of the normal distribution obtained with r=randn(10000); hist(r,m) shown in Figure 11.3 with the histograms of the uniform distribution in Figure 11.1.

If you need to have the same pseudo random number sequence in your calculations, read how to properly use the rng function. This function controls the initial "state" of MATLAB's RNGs.

11.6   Self-Study

Problem 11.1
Consider the LCG random generator with $a = 11$, $c = 2$, $m = 65{,}535$, and $r_1 = 1$. What is the best case scenario for the length or period of the random sequence of this LCG? Estimate the actual length of the non-repeating sequence.

Problem 11.2
Try to estimate the lower bound of the length of the non-repeating sequence for MATLAB's built-in rand generator.

CHAPTER 12

Monte Carlo Simulations

Simulations involving random outcomes are often called *Monte Carlo* simulations. The name comes from Monaco's Monte Carlo area, which has a famous casino featured in many movies and books. Since games of chance (i.e., random outcomes) occur in casinos, it is a fitting name for random number simulations.

This chapter presents several examples of random processes simulations that use the random number generators discussed in Chapter 11. We begin with an example of a ball bouncing on a peg board, discuss a coin flipping simulation, and finally simulate a virus spreading.

12.1 Peg Board

Imagine a ball hitting a board with many layers of pegs. As the ball hits a peg, it might deflect left or right with 50/50 chance. It then travels to a peg in the next layer, where again it can bounce left or right with equal chances, and so on until the ball reaches the last layer. This process is depicted in Figure 12.1. We would like to simulate the whole process with a computer.

We used the word "chance" in the description; therefore, this is clearly a task for the Monte Carlo simulation. The main obstacle is to convert the phrase "50/50" from the human notation to the mathematical form. The phrase indicates that the probability of either case is the same and equal to $50/(50 + 50) = 0.5$. To make a decision about an outcome, we need to generate a random number from the uniform distribution and compare it with 0.5. If it is less then 0.5, we shift the ball position to the left by subtracting 1; otherwise, we shift the ball position to the right by adding 1. Here, we assume that all pegs have spacing of 1. We repeat this process for each peg layer. To generate a random number within the 0–1 interval, we use MATLAB's rand function.

The code that implements this procedure for the given number of balls and the peg board with a given number of layers is shown in Listing 12.1.

Listing 12.1 pegboard.m (available at http://physics.wm.edu/programming_with_MATLAB_book/./ch_monte_carlo_simulations/code/pegboard.m)

```
function [x] = pegboard(Nballs, Nlayers)
%% Calculate a position of the ball after running inside
    the peg board
%       imagine a ball dropped on a nail
%                    o
%                    |
```

```
%               |
%               V
%               *
%              / \   50/50 chance to deflect left or right
%             /   \
%
% now we make a peg board with Nlayers of nails and run
    Nballs
%               *
%              *  *
%             *  *  *
%            *  *  *  *
% the resulting distribution of final balls positions
    should be Gaussian

x=zeros(1,Nballs);
for b=1:Nballs
        for i=1:Nlayers
                if (rand() < .5 )
                        % bounce left
                        x(b)=x(b)-1;
                else
                        % bounce right
                        x(b)=x(b)+1;
                end
        end
end

end
```

Note that the resulting x array holds the final position for each ball. Most of the time, we are not interested in individual outcomes of a Monte Carlo simulation. Instead, we analyze the distribution of outcomes. For the problem at hand, we will make the histogram of the resulting x positions. In our case, we run 10^4 balls (to

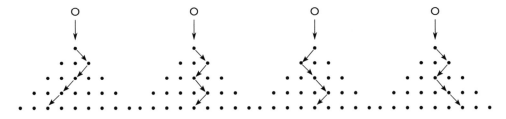

Figure 12.1　Several possible trajectories of the ball bouncing on the peg board.

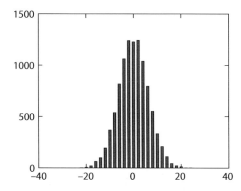

Figure 12.2   The distribution of balls, final positions after 10,000 of them bounced off the 40 layers of pegs.

get a large statistical set) over 40 layers of pegs with the following commands:

```
x=pegboard(10^4, 40);
hist(x, [-25:25]);
```

The resulting distribution of the final x position is shown in Figure 12.2. The balls mostly land in the middle, that is, around $x = 0$, since they bounce left and right during the fall, which tends to self compensate the shifts. The probability of going only to the left (or right) for the 20 bounces is equal to $0.4^{20} \approx 10^{-6}$, which is tiny. This is why the histogram shows almost no counts beyond 20. The resulting distribution resembles a bell shape, that is, it approximates the Gaussian distribution, as it should according to the central limit theorem.

It is left as an exercise for a reader to show that just with simple variable renaming, this code calculates the difference between the numbers of summed heads and tails outcomes for a given number of flips of an unbiased coin.

12.2   Coin Flipping Game

Imagine a game in which someone flips a fair coin. If the coin falls on heads, you receive quadruple your bet; otherwise, you owe your bet to the other player. What fraction of your money should you bet each time to benefit from this game the most, that is, get the maximum gain?

It is easy to see that for each of the coin flips, we should bet the same fraction of money, as the outcome does not depend on the bet amount. We also assume that we can count even a fraction of a penny.

The code in Listing 12.2 shows what might happen if we decide to bet a certain fraction of our capital (bet_fraction) for the given number of games (N_games). The code returns the overall gain after a trip to a casino as the ratio of the final amount of money at hand to the initial one. This time, we use the array calculation capabilities of MATLAB to avoid a loop over each game realization.

Listing 12.2 bet_outcome.m (available at http://physics.wm.edu/
programming_with_MATLAB_book/./ch_monte_carlo_simulations/code/
bet_outcome.m)

```
function gain=bet_outcome(bet_fraction, N_games)
    % We will play a very simple game:
    % one bets a 'bet_fraction' of her belongings
    % with 50/50 chance to win/lose.
    % If lucky she will get her money back quadrupled,
    % otherwise 'bet_fraction' is taken by a casino

    % arguments check
    if (bet_fraction <0 || bet_fraction>1)
        error('bet fraction must be between 0 and 1');
    end
    N_games=floor(N_games);
    if (N_games < 1)
        error('number of games should be bigger than 1');
    end

    p=rand(1,N_games); % get array of random numbers
    outcome_per_game=zeros(1,N_games);

    outcome_per_game(p <= .5) = 1 + 4*bet_fraction; %
        lucky games
    outcome_per_game(p >  .5) = 1 -  bet_fraction; %
        unlucky games

    gain=prod(outcome_per_game);
end
```

We plot the gain vs. the bet fraction for the 200 flips long game in Figure 12.3. The first thing to notice is that the dependence is not smooth. This is natural for random outcomes. Even if we repeat the run with the same bet fraction, the outcome will be different. So, we should not be surprised by the "fuss" (or noise) on our plot. The other thing to notice is that despite the odds being in our favor, it still possible to lose money by the end if the bet fraction is higher than 0.8. Finally, you can see that your gain might be as high as 10^{30} after 200 games if you put in about half of your capital as a bet. This explains why this game is not played in casinos.

12.3 One-Dimensional Infection Spread

In the following simulation, we will make a very simple model of the spread of an infection. We will also see which disease is more dangerous for the population: a

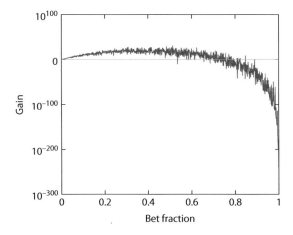

Figure 12.3   The gain of the 200 coin flips game vs. the bet fraction.

highly contagious disease with a low mortality rate or a weakly contagious disease
with a high mortality rate.

We assume that all colony members are in the file arrangement, and each
colony member interacts only with two nearby members: one to the left and one
to the right. The two infections spread only via these interactions.

Our first order of business is to program this interaction. We need a function
that depends on the disease of a particular member and decides whether the infection spreads to the left or right neighbor and also what the self outcome for a given
member or cell is (whether we stay infected, heal, or die). All these outcomes
are probabilistic and depend on certain matrices (prob_to_spread and prob_self
), which store these probabilities for all infections. We treat healthy and dead status
of a member as diseases as well; they simply have zero probability of spreading or
changing their status. Consequently, we assume that a live cell will not die unless
it is infected with one of the deadly diseases. Please study the annotated code in
Listing 12.3, which implements these conditions.

Listing 12.3   disease_outcome.m (available at http://physics.wm.edu/
programming_with_MATLAB_book/./ch_monte_carlo_simulations/code/
disease_outcome.m)

```
function [my_new_infection, infect_left_neighbor,
    infect_right_neighbor] = disease_outcome( disease )
%% For a given disease/infection (defined below) returns
    possible outcomes/actions.
% The disease probabilities matrix:
% notice that probabilities of  self actions   (prob_self)
% i.e. stay ill, die, and heal should add up to 1
% probabilities distributed in a vector where each
    positions corresponds to
die  = 1;
```

```matlab
heal = 2;
% there is no 3rd element since probability of stay ill =
   1 - p_die - p_heal

% array of probability to spread disease (prob_to_spread)
    per cycle
% for each disease
left  = 1;
right = 2;
% the probabilities to infect the left or right neighbors
   are independent,
% the only requirement they should be <=1.

% normal dead member  (death is non contagious)
prob_self(1,:)      = [0.0, 0.0]; % [prob_die, prob_heal]
prob_to_spread(1,:) = [0.0, 0.0]; % [left, right]

% weakly contagious but high mortality, hard to heal
prob_self(2,:)      = [0.8, 0.0];
prob_to_spread(2,:) = [0.1, 0.1];

% highly contagious but low mortality, easy to heal
prob_self(3,:)      = [0.1, 0.1];
prob_to_spread(3,:) = [0.4, 0.4];

% healthy alive member  (life is a disease too)
prob_self(4,:)      = [0.0, 0.0];
prob_to_spread(4,:) = [0.0, 0.0];



%% 1st, do we infect anyone?
% roll the dices for the left neighbor
p=rand();
if ( p  <= prob_to_spread(disease, left) )
    infect_left_neighbor=true;
else
    infect_left_neighbor=false;
end

p=rand(); % reroll the dices for the right neighbor
if( p <= prob_to_spread(disease, right) )
    infect_right_neighbor=true;
else
    infect_right_neighbor=false;
```

```
end

%% 2nd, what is our own fate?
p=rand();
if ( p <= prob_self(disease, die))
    my_new_infection = 1; %death
elseif (p <= (prob_self(disease, die) + prob_self(disease,
    heal)) )
    % notice the sum !
    my_new_infection = 4; %healing, recovery to normal
else
    my_new_infection =  disease; % keep what you have
end
end
```

Now, we are ready to implement the colony evolution. At first, we decide how many members and how many cycles we will track. The one-dimensional array `member_stat` keeps track of the current cycle colony status. We populate it with members that are all healthy (alive); then, we infect a certain number of unlucky cells with either of the infections (there is no check against one infection overriding the other). Then, we begin the evolution cycle, where each cell can die, heal, or stay as before as well as infect the left or the right neighbors. At the end of the cycle, we count the death toll of the given cycle from each of the infections. As the cycles go on, we update the `member_stat_map`, which stores the colony status for every cycle. Please read the annotated code of all of these actions in Listing 12.4.

Listing 12.4 `colony_life.m` (available at `http://physics.wm.edu/` `programming_with_MATLAB_book/./ch_monte_carlo_simulations/code/` `colony_life.m`)

```
Ncycles=50;
Nmembers=200;
Ndiseases=4;

% diseases to number its index translation
death = 1;
hard_to_heal_weakly_contagious = 2;
easy_to_heal_very_contagious = 3; %
alive = 4;




member_stat=zeros(1,Nmembers);
% let's make them all live
member_stat(:)=alive;
```

```
% here we will keep the map of disease spread
member_stat_map=zeros(Ncycles,Nmembers);

% here we will death toll stats for each of the disease
killed_by_disease=zeros(Ncycles,Ndiseases); % so far no
    one is killed

% let's infect a few unlucky individuals
Ndiseased_hard_to_heal_weakly_contagious =20;
for i=1:Ndiseased_hard_to_heal_weakly_contagious
        m=ceil(Nmembers*rand()); % which member in the
            array
        member_stat(m)=hard_to_heal_weakly_contagious;
end
% note that below loop might overwrite one disease with
    another.
Ndiseased_easy_to_heal_very_contagious=20;
for i=1:Ndiseased_easy_to_heal_very_contagious
        m=ceil(Nmembers*rand());
        member_stat(m)=easy_to_heal_very_contagious;
end

% day one stats assignment
member_stat_map(1,:) = member_stat; % first day situation
    recorded

for c=2:Ncycles % on cycle one we just initialize the
    colony
        if c~=1
                killed_by_disease(c,:)=killed_by_disease(c
                -1,:); % accumulative count
        end
        % spread diseases
        for i=1:Nmembers
                disease = member_stat(i);

                [self_acting_disease, ...
                infect_left_neighbor, ...
                infect_right_neighbor] = disease_outcome(
                    disease);

                if (i-1 >= 1)
                        % we have left neighbor
                        if ( infect_left_neighbor == true)
```

```
                               if(member_stat(i-1) ~=
                                  death)
                                        % only alive guys
                                          can catch a
                                          disease
                                        member_stat(i-1)=
                                          disease;
                               end
                       end
               end

               if (i+1 <= Nmembers)
                       % we have right neighbor
                       if ( infect_right_neighbor == true
                          )
                               if(member_stat(i+1) ~=
                                  death)
                                        % only alive guys
                                          can catch a
                                          disease
                                        member_stat(i+1)=
                                          disease;
                               end
                       end
               end

               if ( (self_acting_disease == death) && (
                  disease ~=death) ) % we should not
                  count already dead
                       % add to death toll
                       killed_by_disease(c,disease)=
                          killed_by_disease(c,disease)+1;
               end
               member_stat(i)=self_acting_disease;
       end

       % update member stat vs day map
       member_stat_map(c,:) = member_stat;

end
```

Once we run the `colony_life` command, we are ready to make some plots. At first, we plot the colony evolution map with the following commands:

Listing 12.5 `colony_map_plot.m` (available at http://physics.wm.edu/programming_with_MATLAB_book/./ch_monte_carlo_simulations/code/colony_map_plot.m)

```
% plot the map of the colony evolution
colony_color_scheme=[ ...
        % color coded as RGB triplet
        0,0,0; % color 1, black is for dead
        0,0,1; % color 2, blue is for weakly contagious
        1,0,0; % color 3, red is for highly contagious
        0,1,0; % color 4, green is for healthy
        ];
image( member_stat_map );
set(gca,'FontSize',20); % font increase
colormap(colony_color_scheme);
xlabel('Member position');
ylabel('Cycle');
```

This produces Figure 12.4. We can see that the cells with the deadly disease (light gray color) and a low spread probability die in the first couple of days without infecting their neighbors. The highly infectious cells (dark gray color) start to spread disease left and right while mostly being alive. The borders for the infection spread are set by the clusters of the dead cells (black color). Since the dead cells do not interact, they stop the infection spreading. After about 40 cycles, all cells with the low mortality rate infection die as well.

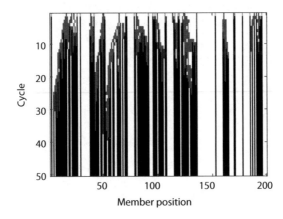

Figure 12.4   The colony evolution map. Rows correspond to evolution cycle; columns reflect the position of a colony member. The white represents a healthy and uninfected colony member, the light gray represents members infected with the hard to heal disease, the dark gray represents members infected with the the highly infectious but low mortality disease, and the black represents dead members.

Now, we execute the commands in Listing 12.6 to make the plot of the accumulated death toll for each cycle for both diseases.

Listing 12.6 `colony_life_death_toll_plot.m` (available at `http://physics.wm.edu/programming_with_MATLAB_book/./ch_monte_carlo_simulations/code/colony_life_death_toll_plot.m`)

```
% only real illness counts
bar(killed_by_disease(:,[hard_to_heal_weakly_contagious,
    easy_to_heal_very_contagious]),'stacked'); % notice the
    array slicing
ylim([0,150]);
set(gca,'FontSize',20); % font increase
legend('death by hard to heal but not contagious', 'death
    by easy to heal but highly contagious');
xlabel('Cycle number');
ylabel('Death toll');
```

These commands produce Figure 12.5. It supports our initial observations: the deadly disease kills all 20 cells initially infected with it within 4 cycles. The not so deadly disease kills more and more as infection spreads, until cycle 40, when everybody with these infection is dead. At the end, we see that high mortality infection kills only 20 members of the colony, while highly contagious disease with low mortality kills about 120 members out of the initial population of 200. So, we can see that a simple flu might be more dangerous than the Ebola virus. Of course, our model is very rudimentary and does not take into account medical facilities, which increase the survival rate drastically.

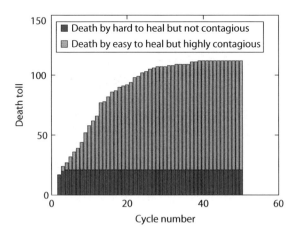

Figure 12.5   The accumulated death toll by two different infections.

12.4   Self-Study

Problem 12.1

Modify the code for the coin flipping game described in Section 12.2 to plot the average of at least 1000 games for each bet fraction. Provide a better estimate for the best bet fraction maximizing your gain.

Problem 12.2

Modify the code for the coin flipping game described in Section 12.2 to take a variable win_for_lucky_flip, which sets the multiplier for the lucky "face up" outcome. Plot the average of at least 1000 games for each bet fraction. Provide an estimate for the best bet fraction maximizing your gain when win_for_lucky_flip=6.

Problem 12.3

Modify the code for the coin flipping game described in Section 12.2 to replace the coin with a six-sided die (with equal probability of landing on any side). The sides of the die are labeled 1, 2, 3, 4, 5, and 6. Your gain for each throw is equal to the die number minus 1. If a 1 is rolled, your bet is taken away by another player. Plot the average of at least 1000 games for each bet fraction. Provide an estimate for the best bet fraction maximizing your gain.

Problem 12.4

Modify the colony life script (see Listing 12.4) to take helpful neighbors into account. Leave the probability of healing a cell without live neighbors as it is, but double the healing probability for a cell with one alive neighbor and triple it if a cell has alive neighbors on both sides. We assume that illness does not prevent a helping/healing action, that is, a neighbor must be alive to help, but it might have an illness too.

CHAPTER 13

The Optimization Problem

This chapter discusses the optimization problem and several approaches to its solution. It discusses optimization in one dimension and several dimensions, then shows MATLAB's built-in optimization commands, and covers combinatorial optimization. The chapter shows two methods that are inspired by nature: simulated annealing and genetic algorithm.

Optimization problems are abundant in our daily lives, as we each have a set of goals and finite resources, which should be optimally allocated. Of all these resources, time is always in demand, as there are natural bounds on our time resources. We are limited to 24 hours in a day, yet we need to allocate time for sleep, studies, work, rest, and multiple other tasks. We are always facing a question: should we sleep an extra hour to be rested before work, or should we instead read an interesting book? Everyone has a different solution, but the problem is the same: how to optimize the distribution of the available resources to get a maximum outcome. In this chapter, we will cover several typical optimization problems and common methods to solve them. But it should be said up front: **there is no guaranteed way to find the global optimal point, that is, the very best, at finite time in a general case.**

13.1 Introduction to Optimization

Before we begin, we will start with a formal mathematical definition of the optimization problem.

The optimization problem

Find $\vec{x}$ that minimizes $E(\vec{x})$ subject to $g(\vec{x}) = 0$ and $h(\vec{x}) \leq 0$,
  where:

$\vec{x}$ is the vector of independent variables

$E(\vec{x})$ is the energy function, which sometimes is also called the *objective* or *fitness* or *merit function*

$g(\vec{x})$ and $h(\vec{x})$ are constraining functions

As you can see, we solve the optimization toward a minimum, that is, the minimization problem in computer science. It is easy to see that maximization problems are the same as minimization once we change $E(\vec{x}) \rightarrow -E(\vec{x})$.

For a physicist, it is clear why the minimized function is called *energy*. We are looking for the minimum, or lowest point, which is the point on the potential energy landscape where a physical system tends to arrive (in the presence of the dissipative forces), that is, nature constantly solves the minimization of energy problem.

Figure 13.1   Example of the function to be minimized.

The constraining functions are dealing with some additional dependencies among the components of the $\vec{x}$. If we have a budget of \$ 100 per day, we allocate certain amounts for food (x_1), books (x_2), movies (x_3), and clothes (x_4). Our final goal is to maximize overall happiness or, since we are doing minimization, to minimize unhappiness. Everyone has a different merit function, which depends on the above parameters, although there is an obvious common constraint: we cannot spend more than \$100, so $h(\vec{x}) = (x_1 + x_2 + x_3 + x_4) - 100 \leq 0$. The constraining function can take the form of conditional statements or set a limit for only a few parameters. For example, we can say that we need to spend at least some amount of money on food. There are also unconstrained problems, in which any value of $\vec{x}$ is permitted.

13.2   One-Dimensional Optimization

At first, we consider the one-dimensional optimization problem, that is, the ($\vec{x}$) has only one component. In this case, we can drop the vector notation and just use x for the independent variable, as x is one-dimensional. Suppose that dependence of our merit or energy function (E) on x looks as shown in Figure 13.1. If we know the analytical expression for $E(x)$, we can find an expression for its derivative $f(x) = dE/dx$. Then, we find the positions of extrema by solving $f(x) = 0$ and checking which of them belongs to the global minimum (recall that some of them might belong to local minima or even maxima). We just reduced the optimization problem to the root finding problem, for which we have a variety of solution algorithms, discussed in Chapter 8.

Interestingly enough, if we know how to solve the minimization problem, we can use it for the root finding problem, that is, $f(x) = 0$. This is done by assigning the merit function to be $E(x) = f(x)^2$. Since such $E(x) \geq 0$, the global minimum position (x_m) where $E = 0$ coincides with the root of $f(x)$.

Now, let's discuss the general requirements for the minimization algorithm. We need to somehow bracket the minimum (optimum) location and then iteratively reduce the bracket size until we find the precision of the optimum location (given by the bracket length) satisfactory. Let's assume that we are in *the close vicinity of a minimum*, that is, there is no other extremum inside the bracket. If you think a bit about the problem, you will see that probing only one test point within the bracket does not provide enough information to correctly assign the new ends of the bracket. So, we need to calculate the function value for at least two inner points; then, we can assign the new bracket by the following rule: the bracket ends should be the two closest points (among already checked: bracket ends and two inner points) surrounding the lowest currently known merit point. Then, we can repeat this bracket updating procedure until the required precision is reached.

Now, the key question becomes: how do we choose two inner points for the above general algorithm efficiently*? Quite often, the merit function is expensive to calculate either in terms of the time required for calculation or sometimes literally (if you optimize a rocket engine, it is not cheap to build a new test engine). Therefore, we would like to reduce the number of merit function calculations per bracket reduction.

13.2.1 The golden section optimum search algorithm

The golden section optimum search algorithm addresses the efficiency question by reusing one of the two previous test points and, consequently, requires only one additional function calculation per bracket update.

The golden section optimization algorithm

Assign a bracket interval (a, b) that surrounds a minimum (ideally the global minimum) closely enough, that is, there are no other extrema in the interval (this is similar to the requirement in item 4).

1. Calculate $h = (b - a)$.

2. Assign new probe points $x_1 = a + R \times h$ and $x_2 = b - R \times h$, where

$$R = \frac{3 - \sqrt{5}}{2} \approx 0.38197 \tag{13.1}$$

3. Calculate $E_1 = E(x_1)$, $E_2 = E(x_2)$, $E_a = E(a)$, and $E_b = E(b)$.

4. We require the bracket size to be small enough: h: $E(x_1) \leq E(a)$ and $E(x_2) \leq E(b)$. **This is important!** In this case, we can shrink or update the bracket:

   - If $E_1 < E_2$, then $b = x_2$ and $E_b = E_2$, else $a = x_1$ and $E_a = E_1$.

   - Recalculate $h = (b - a)$.

5. If the required precision is reached, that is, $h < \varepsilon_x$, then stop (choose any point within the bracket for the final answer); otherwise, do the following steps.

6. Reuse one of the old points, either (x_1, E_1) or (x_2, E_2).

   - If $E_1 < E_2$,
     then $x_2 = x_1$, $E_2 = E_1$, $x_1 = a + R \times h$, $E_1 = E(x_1)$,
     else $x_1 = x_2$, $E_1 = E_2$, $x_2 = b - R \times h$, $E_2 = E(x_2)$.

7. Repeat from step 4.

---

* Efficiently means optimally, so we are solving yet another optimization problem.

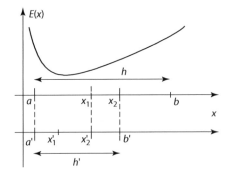

Figure 13.2   The golden section minimum search method illustration.

It is easy to see that the new bracket size $h' = (1 - R) \times h$. In other words, the bracket shrinks by the factor $1 - R = \varphi = (\sqrt{5} - 1)/2 \approx 0.61803$, which is the golden ratio.* This gives its name to the whole algorithm.

13.2.1.1   Derivation of the R coefficient
Let's derive the expression for the R coefficient. Look at the golden section optimization algorithm; at the first step, we have

$$x_1 = a + R \times h \tag{13.2}$$
$$x_2 = b - R \times h \tag{13.3}$$

As depicted in Figure 13.2, if $E(x_1) < E(x_2)$, $a' = a$, and $b' = x_2$, then we assign the next probing points x_1' and x_2' according to

$$x_1' = a' + R \times h' = a' + R \times (b' - a') \tag{13.4}$$
$$x_2' = b' - R \times h' = b' - R \times (b' - a') = x_2 - R \times (x_2 - a) \tag{13.5}$$

We would like to reuse one of the previous evaluations of E, so we require that $x_1 = x_2'$. Using this, we plug Equation 13.2 into the right-hand side of Equation 13.5 and obtain

$$a + R \times h = b - R \times h - R \times (b - R \times h - a) \tag{13.6}$$

---

* The golden ratio dates from the time of the ancient Greeks and naturally comes up in the solutions of several geometrical and mathematical problems. Some even argue that it has some aesthetic properties for geometrical constructs. For example, a rectangle with the ratio of the short side to the long side equal to φ is apparently pleasing to the eyes. Maybe this is why more and more computer screens have sides of ratio 9 to 16, which is close to φ.

Recalling that $h = b - a$, we can cancel it out and obtain the quadratic equation

$$R^2 - 3R + 1 = 0$$

with two possible roots:

$$R = \frac{3 \pm \sqrt{5}}{2}$$

We need to choose the solution with the minus sign, since $R \times h$ should be less than h to land the two probe points inside the bracket.

13.2.2 MATLAB's built-in function for the one-dimension optimization

MATLAB has a built-in function `fminbnd` to perform the one-dimension optimization, which uses the modified golden section search algorithm for its implementation. The `fminbnd` function takes three mandatory arguments: the handle to the merit function, the left end of the bracket, and the right end of the bracket. We can also supply some additional options that, for example, set the required precision for the minimum location or the number of the permitted function evaluations.

13.2.3 One-dimensional optimization examples

13.2.3.1 Maximum of the black body radiation

In physics, an object is called a *black body* if it does not reflect electro-magnetic radiation. Surprisingly, a black body could radiate quite a lot of energy and thus appear bright when it is hot enough. In this regard, our sun is actually an almost perfect black body, and so is an incandescent bulb when it is on.

According to Plank's law, the spectrum of the power radiated per area of the black body per wavelength into the solid angle has the following dependence on wavelength of electro-magnetic radiation (λ) and temperature (T):

$$I(\lambda, T) = \frac{2hc^2}{\lambda^5} \frac{1}{e^{\frac{hc}{\lambda kT}} - 1} \tag{13.7}$$

where:
    h is the Planck constant 6.626×10^{-34} J$\times$s
    c is the speed of light 2.998×10^8 m/s
    k is the Boltzmann constant 1.380×10^{-23} J/K
    T is the body temperature in K
    λ is the wavelength in meters

For an incandescent bulb with a typical filament temperature of 1500 K, the black body radiation spectrum looks as depicted in Figure 13.3. We calculate it with the help of the function, which implements Equation 13.7, in Listing 13.1.

Listing 13.1 `black_body_radiation.m` (available at `http://physics.wm.edu/programming_with_MATLAB_book/./ch_optimization/code/black_body_radiation.m`)

```
function I_lambda=black_body_radiation(lambda,T)
% black body radiation spectrum
% lambda - wavelength of EM wave
% T - temperature of a black body
h=6.626e-34;   % the Plank constant
c=2.998e8;     % the speed of light
k=1.380e-23;   % the Boltzmann constant

I_lambda = 2*h*c^2 ./ (lambda.^5)  ./ (exp(h*c./(lambda*k*
    T))-1);
end
```

It is easy to see that most of the radiation is emitted above the 1000 nm wavelength, where the human eye has no ability to register the light. Consequently, incandescent bulbs are not very efficient at providing light, since most of the energy becomes heat (i.e., infrared radiation). It's no wonder there is a big effort to replace incandescent bulbs with modern fluorescent or LED bulbs, which provide much more efficient lighting.

Suppose we would like to know the wavelength of the sun's maximum radiation. MATLAB knows how to find a minimum of the function, so we create the merit function f, which is black_body_radiation reflected (inverted) with respect to the x-axis. The sun's photosphere temperature is 5778 K, so we set the T accordingly.

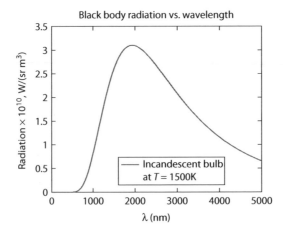

Figure 13.3   The black body radiation spectrum for an incandescent bulb with the filament temperature T = 1500 K.

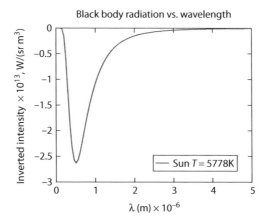

Figure 13.4 The inverted radiation spectrum of the sun or any black body with the temperature T = 5778 K.

```
T=5778;
f = @(x) - black_body_radiation(x,T);
```

The resulting plot of the merit function, that is, the inverted black body radiation spectrum of the sun, is depicted in Figure 13.4. As we can see, the minimum is located somewhere in the $(10^{-9}, 10^{-6})$ interval. We need to adjust the default x tolerances, since the typical x value, as shown in Figure 13.4 is in the order of 10^{-6}. which is MATLAB's default precision. We use the optimset command to tune the precision. Now, we are ready to search the minimum location with the following commands:

```
fminbnd(f, 1e-9, 1e-6, optimset('TolX',1e-12))
ans = 5.0176e-07
```

As you can see, the answer is 5.0176e—07, measured in meters, so the maximum of the sun's radiation is at roughly 502 nm, which corresponds to green light.* No wonder that the human eye is most sensitive to green light, since it is the dominating wavelength in a naturally lit environment.

13.3 Multidimensional Optimization

We will not talk about algorithms of multidimensional optimization for smooth functions here.[†] If you are interested in them, have a look at specialized numerical methods books, for example, [9]. Instead, we will piggyback on MATLAB's fminsearch function.

---

* Strangely enough, the Sun appears to be yellowish to a human eye. This is due to a particular response of the light-sensitive elements in the eye and in the brain that reconstruct the perceived color. One side effect is that there are white, blue, yellow, and red stars (which are all black bodies listed in the order of decreasing temperature), but there are no green stars in the sky.

† It is the author's opinion that programming such algorithms do not bring much educational value. MATLAB has good enough and ready to use implementations.

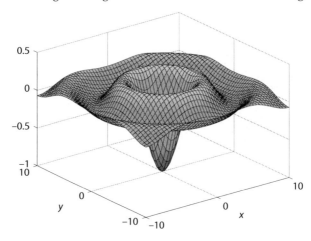

Figure 13.5   The two-dimension sinc function plot.

13.3.1   Examples of multidimensional optimization

13.3.1.1   The inversed sinc function
Let's find a minimum of the inverse two-dimension sinc function

$$f1(x,y) = -\sin(r)/r, \text{ where } r = \sqrt{x^2 + y^2} \tag{13.8}$$

The plot of this function is shown in Figure 13.5.

To do the optimization with fminsearch, we need to implement Equation 13.8 as a function of only one vector argument, as shown in Listing 13.2.

Listing 13.2   fsample_sinc.m (available at
http://physics.wm.edu/programming_with_MATLAB_book/./
ch_optimization/code/fsample_sinc.m)

```
function ret=fsample_sinc(v)
        x=v(1); y=v(2);
        r=sqrt(x^2+y^2);
        ret= -sin(r)/r;
end
```

The components of the input vector v are representing x and y coordinates. It is up to us to assign x as the first component and y as the second one; we can do it the other way as well.

To call fminsearch, we need two arguments: the first is the handle to the function to be optimized (i.e., @fsample_sinc), and the second is a starting point for the minimum search algorithm (in this example, we will use [0.5, 0.5]). Once we have made this decision, we are ready to search for the minimum:

```
>> x0vec=[0.5, 0.5];
>> [x_opt,z_opt]=fminsearch(@fsample_sinc, x0vec)
   x_opt = [0.2852e-4,    0.1043e-4]
   z_opt = -1.0000
```

As we can see, the minimum is located at x_opt=[0.2852e−4, 0.1043e−4], which is very close to the true global minimum at [0,0]. The value of the function in the found optimum location is z_opt = −1.0000, which matches (within the shown precision) the global minimum value −1.

It is easy to miss the global minimum if we choose a bad starting point, as in this example:

```
>> x0vec=[5, 5];
>> [x_opt,z_opt]=fminsearch(@fsample_sinc, x0vec)
   x_opt = [ 5.6560    5.2621 ]
   z_opt = -0.1284
```

Here, we find a local minimum but not the global minimum. Recall that no algorithm can find the global minimum in a general case, especially when it starts far away from the optimum.

13.3.1.2 Three-dimensional optimization
Let's find the minimum of the function

$$f2(x,y,z) = 2x^2 + y^2 + 2z^2 + 2xy + 1 - 2z + 2xz. \tag{13.9}$$

We do it by implementing $f2$ as shown in Listing 13.3.

Listing 13.3 f2.m (available at http://physics.wm.edu/
programming_with_MATLAB_book/./ch_optimization/code/f2.m)

```
function fval = f2( v )
x = v(1);
y = v(2);
z = v(3);
fval = 2*x^2+y^2+2*z^2+2*x*y+1-2*z+2*x*z;
end
```

Yet again, it is up to us which component of the input vector we use as x, y, and z. To find the minimum, we choose an arbitrary starting point [1,2,3] and execute

```
>> [v_opt, f2_opt]=fminsearch(@f2, [1,2,3])
v_opt = -1.0000    1.0000    1.0000
f2_opt = 4.8280e-10
```

At first glance, it may not be clear how to check the calculated minimum position v_opt = -1.0000 1.0000 1.0000, but we can rewrite Equation 13.9:

$$f2(x,y,z) = (x+y)^2 + (x+z)^2 + (z-1)^2 \tag{13.10}$$

Since every term is quadratic, the minimum is reached when each term is equal to zero. So, the global minimum is at $[x,y,z] = [-1,1,1]$. For the same reason, $f2$ cannot be less than zero; so, f2_opt = 4.8280e-10, which is very close to zero, is appropriate.

13.3.1.3   Joining two functions smoothly

Suppose we have a function that has the following form:[†]

$$\Psi(x) = \begin{cases} \Psi_{in}(x) = \sin(kx) & : \quad 0 \le x \le L \\ \Psi_{out}(x) = Be^{-\alpha x} & : \qquad x > L \end{cases}$$

We would like to make our function smooth, that is, both the function and its first derivative are continuous everywhere. The only problem point is located at $x = L$, where one continuous and smooth function meets another. The following equations are in charge of the smooth link conditions:

$$\Psi_{in}(L) = \Psi_{out}(L) \tag{13.11}$$

$$\Psi'_{in}(L) = \Psi'_{out}(L) \tag{13.12}$$

After substitution of the Ψ expression, we obtain

$$\sin(kL) = Be^{-\alpha L} \tag{13.13}$$

$$k\cos(kL) = -\alpha Be^{-\alpha L} \tag{13.14}$$

Suppose that we somehow know k. What should be the values of α and B? We can solve this system of nonlinear equations to get α and B, but this is a tedious task. Besides, this chapter is about optimization. So, we will use our new skills to solve

---

[†] You are probably interested in where this function comes from. This is the solution of the quantum mechanics problem about a particle in a one-dimensional potential well described by the following potential:

$$U(x) = \begin{cases} \infty & : \quad x < 0 \\ 0 & : \quad 0 \le x \le L \\ U_o & : \qquad x > L \end{cases}$$

where:

$k = \dfrac{\sqrt{2m(E-U_o)}}{\hbar}$, $\alpha = \dfrac{\sqrt{2m(U_o-E)}}{\hbar}$, m is the mass of the particle

E is its total energy

$\hbar = h/(2\pi)$ is the reduced Planck constant

Since the potential is infinite at $x < 0$, $\Psi(x) = 0$ in this region.

the problem. We rearrange the equations as

$$\sin(kL) - Be^{-\alpha L} = 0 \tag{13.15}$$

$$k\cos(kL) + \alpha Be^{-\alpha L} = 0 \tag{13.16}$$

then we square and add them together:

$$\left(\sin(kL) - Be^{-\alpha L}\right)^2 + \left(k\cos(kL) + \alpha Be^{-\alpha L}\right)^2 = 0. \tag{13.17}$$

So far, we have not done anything out of the ordinary. Now, we call the right-hand side of the above equation as the merit of our problem:

$$M(\alpha, B) = \left(\sin(kL) - Be^{-\alpha L}\right)^2 + \left(k\cos(kL) + \alpha Be^{-\alpha L}\right)^2. \tag{13.18}$$

The global minimum of the merit function is the point in α and B space where Equations (13.15) and (13.16) are satisfied. The listing of the merit function is shown in Listing 13.4.

Listing 13.4 `merit_psi.m` (available at `http://physics.wm.edu/programming_with_MATLAB_book/./ch_optimization/code/merit_psi.m`)

```
function [m] = merit_psi(v, k , L)
% merit for the potential well problem
alpha=v(1);
B=v(2);

m=(sin(k*L)  - B*exp(-alpha*L))^2 + (k*cos(k*L)  + alpha*B*
    exp(-alpha*L))^2;

end
```

All we need to do is assign the k and L values and make the `fminsearch` compatible merit function (i.e., the one that accepts the problem parameters vector). All of this is done by executing the following code:

```
>> k=2+pi; L=1;
>> merit=@(v) merit_psi(v, k, L);
>> v0=fminsearch( @merit, [.11,1] )
v0 = 2.3531    -9.5640
```

The resulting values are $\alpha = 2.3531$ and $B = -9.5640$. The plot of the Ψ function with these values is shown in Figure 13.6. As you can see, the transition between inner and outer parts of the Ψ function is smooth, as required.

13.3.1.4 Hanging weights problem

Consider the masses m_1 and m_2, which are connected by rods with length L_1, L_2, and L_3 to suspension points hanging in Earth's gravitational field (see Figure 13.7).

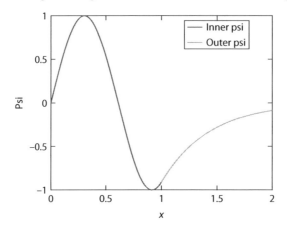

Figure 13.6   The plot of the smoothly connected inner and outer parts of the Ψ function.

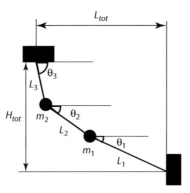

Figure 13.7   The suspended weights arrangement.

The suspension points are separated horizontally by L_{tot} and vertically by H_{tot} distances. Our goal is to find the angles θ_1, θ_2, and θ_3 of these weights arrangement in equilibrium. This is a typical Physics 101 problem. To solve it, we need to set and solve several equations regarding forces and torques acting on the system. This is not a trivial task. You might wonder what this problem has to do with optimization. You will soon see that this problem can be replaced with the minimization problem and solved quite elegantly (i.e., with fewer equations to track). The downside is that the solution will be numerical, that is, we will have to redo the calculations if some parameters change.

Recall that the other name of the merit function is energy, for a quite important reason: a real life system seeks the minimum of the potential energy due to forces of nature. So, we need to minimize the potential energy subject to the length constraints. The latter requirement is important, since the potential energy minimization alone will push our weights to the lowest position, but they are connected by the links, so this needs to be taken into account. See the code in Listing 13.5 for the resulting merit function of this problem. Note that we already entered the particular values for the masses, lengths of each rod, and suspension point separation.

Listing 13.5 `EconstrainedSuspendedWeights.m` (available at `http://physics.wm.edu/programming_with_MATLAB_book/./ch_optimization/code/EconstrainedSuspendedWeights.m`)

```
function [merit, LengthMismatchPenalty,
    HeightMismatchPenalty] = EconstrainedSuspendedWeights(
    v )
% reassign input vector elements to the meaningful
    variables
theta1=v(1); theta2=v(2); theta3=v(3); % theta angles

g=9.8; % acceleration due to gravity
m1=2; m2=2; % masses of the weights
L1=3; L2=2;  L3=3; % lengths of each rod
Ltot=4; Htot=0; % suspension points separations
% fudge coefficients to make merit of the potential energy
    comparable to
% length mismatch
mu1=1000; mu2=1000;

Upot=g*( (m1+m2)*L1*sin(theta1)+m2*L2*sin(theta2) ); %
    potential energy
HeightMismatchPenalty=(Htot-(L1*sin(theta1)+L2*sin(theta2)
    +L3*sin(theta3)))^2;
LengthMismatchPenalty=(Ltot-(L1*cos(theta1)+L2*cos(theta2)
    +L3*cos(theta3)))^2;

merit=Upot+mu1*LengthMismatchPenalty+mu2*
    HeightMismatchPenalty;
end
```

This code needs some walkthrough. Why is length mismatch called a penalty? If we have a length mismatch for some test point, we need to let the solver know that we are breaking some constraints, that is, we need to penalize such a probe point. Since we are looking for the minimum, we add something positive to the resulting merit value at this point. It is usually a good idea to add a square of mismatch: it is always positive and smooth. The `mu1` and `mu2` coefficients emphasize the importance of the particular contribution to the resulting merit. Their assignment usually requires some tuning to make all contributions equally important.

For the problem with the chosen parameters, that is,

```
m1=2; m2=2;
L1=3; L2=2;  L3=3;
Ltot=4; Htot=0;
```

we can notice the symmetry: the masses are the same, and the outer rods are the same. So, we can be sure that the inner rod (L_2) should be horizontal, that is,

$\theta_2 = 0$; additionally, θ_1 should be equal to $-\theta_3$ due to the same symmetry. We can even find their values precisely: $\theta_1 = -1.231$ and $\theta_3 = 1.231$. Let's see if the minimization algorithm gives the correct answer.

```
>> theta = fminsearch( @EconstrainedSuspendedWeights,
    [-1,0,-1], optimset('TolX',1e-6))
theta = -1.2321    -0.0044    1.2311
```

You can see that the answer is quite close to the theoretically predicted values. So, we declare our approach successful.

13.4 Combinatorial Optimization

There is a subclass of problems for which the parameters vector or its components can take only discrete values. For example, you can only buy hot dog buns in sets of eight. So, when you are optimizing your spending for a party, you would have to account for 0 or 8 or 16 ... as a possible number of buns.

As a result of this discretization, the optimization algorithms and function, which we have covered before, are of little use. They assume that any component can take any value and that the merit function will be fine with it. There is a way around this. We can create constraining functions that take care of it, but this is generally not a trivial task.

Instead, we have to find a method to search through discrete sets of all possible input values, that is, try all possible combinations of $\vec{x}$ components. Hence, the name of the optimum search is *combinatorial optimization*.

Unfortunately, there is no way to design a general optimum searching algorithm that can solve any combinatorial problem. So, every combinatorial problem requires a specific solution, but the general idea is the following: probe every possible combination of the inputs and select the best. Usually, the hardest part is to devise a method to go over all possible combinations (ideally without repeating any that have already been probed).

We will cover two problems that should give you a general idea about how to approach problems of this type.

13.4.1 Backpack problem

Suppose you have a backpack with a given size (volume) and a set of objects with given volumes and monetary (or sentimental) values. Our job is to find the subset of items that can be packed into the backpack and has the maximum combined value. For simplicity, we will assume that every item occurs only once.

The mathematical formulation of this problem is the following: maximize the merit function

$$E(\vec{x}) = \sum \text{value}_i x_i = \overrightarrow{\text{values}} \cdot \vec{x}$$

subject to the following constraint

$$\sum \text{volume}_i x_i = \overrightarrow{\text{volumes}} \cdot \vec{x} \leq \text{BackpackSize}$$

where x_i can be 0 or 1, that is, it reflects whether we pack the ith item or not.

We will try a brute force approach, that is, we will check every possible combination of the items.[*] For each item, there are two possible outcomes (pack or do not pack). If we have N items, the number of all possible combinations is 2^N. So, both the size of all possible combinations (i.e., the problem space) and the solving time grow **exponentially**. On the bright side, we will find the **global** optimum.

The hardest part of the problem is to find a way to generate all possible combinations of objects to leave or take. We note that the vector $\vec{x}$ is a combination of zeros and ones. For example, the vector might be $\vec{x} = [0,1,0,1,\cdots,1,1,0,1,1]$. A combination of zeros and ones resembles a binary number. The set of all zeros corresponds to the smallest possible positive integer number, that is, 0, and the set of all ones corresponds to the largest possible binary number constructed with N ones: $2^N - 1$. There is a simple recipe for generating all possible integer numbers from 0 to $2^N - 1$: start from 0 and just keep adding 1 to get the next one. The tricky part is to do it according to binary arithmetic, or more precisely, to implement the proper tracking of the digit overflow mechanism.[*] All modern computers use the binary system under the hood, but, strangely enough, we would have to put some effort into the proper implementation of it.[†]

The pseudo-code for probing all $\vec{x}$ combinations for N objects would be the following:

Pseudo-code for the backpack problem

1. Start with $\vec{x} = [0,0,0,0,\cdots,0,0]$ consisting of N zeros.

2. Every new $\vec{x}$ will be generated by adding 1 to the previous $\vec{x}$ according to binary addition rules.

   - For example, $x_{next} = [1,0,1,\cdots,1,1,0,1,1] + 1 = [1,0,1,\cdots,1,1,1,0,0]$.

3. For every new $\vec{x}$, check whether the items fit into the backpack and whether the new packed value is larger than the previously found maximally packed value.

4. We are done once we have tried all 2^N combinations of $\vec{x}$.

MATLAB's realization of this algorithm is shown in Listing 13.6.

---

[*] There are better ways. We can be more selective about how we select items to pack. For example, we can presort all items in ascending order and put them one by one; if the current item does not fit, there is no reason to probe even larger items. This will save computational time. Another way is to use the simulated annealing algorithm (see Section 13.5) to find a good enough solution.

[*] In the decimal system, we cannot add 1 to the largest decimal symbol (9) without using an extra digit, that is, $9 + 1 = 10$. Similarly, in the binary system, where the largest symbol is 1, $1 + 1 = 10_2$.

[†] There are MATLAB functions that deal with the conversion to and from binary numbers.

Listing 13.6 `backpack_binary.m` (available at `http://physics.wm.edu/`
`programming_with_MATLAB_book/./ch_optimization/code/`
`backpack_binary.m`)

```
function [items_to_take, max_packed_value,
    max_packed_volume] = ...
    backpack_binary( backpack_size, volumes, values )
% Returns the list of items which fit in backpack and have
    maximum total value
% backpack_size - the total volume of the backpack
% volumes       - the vector of items volumes
% values        - the vector of items values

% We need to generate vector x which holds designation:
    take or do not take
% for each item.
% For example x=[1,0,0,1,1] means take only 1st, 4th, and
    5th items.
% To generate all possible cases, go over all possible
    combos of 1 and 0
% It is easy to see the similarity to the binary number
    presentation.
% We will start with x=[0, 0, 0, ... ,1]
% and add 1 to the last element according to the binary
    arithmetic rules
% until we reach x=[1, 1, 1, ... ,1] and
% then x=[0, 0, 0, ... , 0], which is the overfilled
    [111..1] +1.
% This routine will sample all possible combinations.

% nested function does the analog to the binary 1 addition
function xout=add_one(x)
    xout = x;
    for i=N:-1:1
        xout(i)=x(i)+1;
        if (xout(i) == 1 )
            % We added 1 to 0. There is no overfill, and
                we can stop here.
            break;
        else
            % We added 1 to 1. According to the binary
                arithmetic,
            % it is equal to 10.
            % We need to move the overfilled 1 to the next
                digit.
```

```
              xout(i)=0;
          end
      end
end


% initialization
N=length(values);    % the number of items
xbest=zeros(1,N);    % we start with empty backpack, as the
      current best
max_packed_value=0; % the empty backpack has zero value

x=zeros(1, N); x(end)=1; % assigning 00000..001 the very
    first choice set

while ( any(x~=0) ) % while the combination is not
    [000..000]
      items_volume = sum(volumes .* x);
      items_value  = sum(values  .* x);
      if ( (items_volume <= backpack_size) && (items_value >
          max_packed_value) )
          xbest=x;
          max_packed_value=items_value;
          max_packed_volume=items_volume;
      end
      x=add_one(x);
end

indexes=1:N;
items_to_take=indexes(xbest==1); % converting x in the
    human notation
end
```

The interesting part of this code is the add_one subfunction, which runs binary addition. Another feature is the use of indexes at the next to last line. This returns a human readable list of the objects to pack, instead of $\vec{x}$ consisting of zeros and ones. The rest is just bookkeeping.

We can test the backpack algorithm with a list of five items of various values and volumes.

```
>> backpack_size=7;
>> volumes=[  2,  5,  1,  3, 3];
>> values =[ 10, 12, 23, 45, 4];
>> [items_to_take, max_packed_value] = ...
        backpack_binary( backpack_size, volumes, values)
  items_to_take = [1 3 4]
  max_packed_value = 78
```

As you can see, the algorithm suggests taking the first, third, and fourth items to maximize the total packed value. There is no better solution than this, as we can see by solving this problem ourselves.

The algorithm searches through all combinations of five objects almost instantaneously. To go over the list of 20 items, my computer takes 24 seconds. It would take almost 1000 times longer to sort through 30 items, that is, more than 6 hours. It is unpractical to use this algorithm to sort through even a slightly longer list of objects. This is the price for the ability to find the global optimum via probing all 2^N combinations.

Words of wisdom

Use of brute force algorithms is never a good idea: they are fast to implement and slow to use.

13.4.2   Traveling salesman problem

Suppose that a salesman has a list of N cities with given coordinates (x and y) to visit. The salesman starts in the city labeled 1 and needs to be in the Nth city at the end of a route (see Figure 13.8). We need to find the shortest route so that the salesman visits every city and does it only once.

This problem has many connections to the real world. Every time you ask your navigator to find a route from one place to another, the navigator unit has to solve a very similar problem. However, the navigator has to select intermediate locations and then find the shortest route. If you choose the destination too far away, the navigator may even complain that it does not have enough resources to do the planning and may suggest choosing an intermediate destination. In the following, you will see why planning a long route with too many places to visit is a hard problem for a computer (at least if a brute force approach is taken).

Let's estimate the problem size of our traveling salesman problem, that is, how many possible combinations exist. If we have N cities in total, the salesman can go from the first city to $N - 2$ destinations. We subtract 2 because the first and last cities are predefined by the problem. For the third city to visit, we have $N - 3$

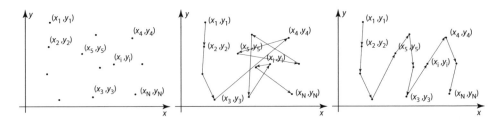

Figure 13.8   The traveling salesman problem illustration. The N cities arrangement is shown at the left; a possible sub-optimal route is shown in the middle; a shorter route is shown at the right.

choices. For the fourth city, we have $N - 4$ cities. The sequence goes on until we have no choices. So, the total number of choices is given by

$$(N - 2) \times (N - 3) \times (N - 4) \times \cdots \times 2 \times 1 = (N - 2)! \qquad (13.19)$$

This grows even faster than exponential dependence. Recall Stirling's approximation: $N! \sim \sqrt{2\pi N}(N/e)^N$. If we spend only a nanosecond to test every route consisting of 22 cities, it would take about 77 years to go over all possible combinations, since $20! \approx 2.4 \times 10^{18}$. Now you see why choosing a route is quite a hard problem for a navigator.[†]

Let's not be discouraged by these numbers. We will be able to select the shortest route from 10 cities within a minute even if we go over all combinations. As before, the hard part is to find a way to go through all possible permitted combinations of the cities. We do this by noticing that a complete route involves all cities, so another route could be achieved by swapping positions of any two cities in the route assignment, that is, by a permutation. So, we need to find a way to go over all possible permutations.

13.4.2.1 Permutation generating algorithm
Luckily, there are permutation generating algorithms available. MATLAB has one of them, implemented as the perms function. Unfortunately, it is not suitable for our needs, since it generates and stores a list of **all** permutations. This consumes all available computer memory even for a modest $N \approx 15$. Instead, we will use a method that goes back to fourteenth-century India (see "Generating all permutations" in [7]). The pseudo-code of this algorithm is shown in the following box.

Next lexicographic permutation generating algorithm

1. Start with the set sorted in ascending order, that is, $p = [1, 2, 3, 4, \cdots, N - 2, N - 1, N]$
2. Find the largest index k such that $p(k) < p(k + 1)$.
   - If no such index exists, the permutation is the last permutation.
3. Find the largest index l such that $p(k) < p(l)$.
   - There is at least one $l = k + 1$.
4. Swap $p(k)$ with $p(l)$.
5. Reverse the sequence from $p(k + 1)$ up to and including the final element $p(end)$.
6. We have a new permutation. If we need another, repeat from step 2.

[†] As with the backpack problem, there are better algorithms. Some are smart about ruling out suboptimal routes without even testing them; they still find the global minimum but with fewer tests. For example, the Held–Karp algorithm does it in $O(2^N N^2)$ steps [3], though this one requires quite a lot of memory to work. Other algorithms find a *good enough* route. An example is the simulated annealing algorithm, which we will see very soon in Section 13.5.

To generate a new permutation, it only needs to know the previous one, so the algorithm memory footprint is negligible. The name *lexicographic* comes from the requirement of the items to be sortable (i.e., we can compare their values). In the past, they used letters, since there is a particular order (i.e., ranking) of them in the alphabet. We do not have to use letters, since numbers naturally possess this property. See MATLAB's implementation of this algorithm in Listing 13.7.

Listing 13.7 `permutation.m` (available at `http://physics.wm.edu/programming_with_MATLAB_book/./ch_optimization/code/permutation.m`)

```
function pnew=permutation(p)
        % Generates a new permutation from the old one
        % in such a way that new one will be
          lexicographically larger.
        %
        % If one wants all possible permutations, she
        % must prearrange elements of the permutation
          vector p
        % in ascending order for the first input, and then
    % feed the output of this function to itself.
        %
        % Elements of the input vector allowed to be not
          unique.
        %
        % See "The Art of Computer Programming, Volume 4:
        % Generating All Tuples and Permutations" by
          Donald Knuth
        % for the discussion of the algorithm.
        %
        % This implementation is optimized for MATLAB. It
          avoids cycles
        % which are costly during execution.

        N=length(p);
        indxs=1:N; % indexes of permutation elements

        % looking for the largest k where p(k) < p(k+1)
        k_candidates=indxs( p(1:N-1) < p(2:N) );
        if ( isempty(k_candidates) )
                % No such k is found thus nothing to
                   permute.
                pnew= p;
        % We must check at the caller for this special
          case pnew==p
        % as condition to stop.
```

```
            % All possible permutations are probed by this
                point.
            return;
        end
        k=k_candidates(end); % note special operator 'end'
                the last element of array

            % Assign the largest l such that p(k) < p(l).
            % Since we are  here at least one solution is
                possible: l= k+1
            indxs=indxs(k+1:end); % we need to truncate the
                list of possible indexes
            l_candidates=indxs( p(k) < p (k+1:end) );
            l=l_candidates(end);

            tmp=p(l); p(l)=p(k); p(k)=tmp; % swap p(k) and p(l
                )

            %reverse the sequence between p(k+1) and p(end)
            p(k+1:end)=p( end:-1:k+1 );
            pnew=p;
end
```

The important thing to mention about this code is that once the last permutation is reached (all items will be sorted in descending order), it will output the same combination that was the input. It is up to us to check for this condition to stop the search.

13.4.2.2   Combinatorial solution of the traveling salesman problem

Once we have a permutation generating algorithm, the rest is straightforward bookkeeping to find the shortest route. MATLAB's solution of the problem is shown in Listing 13.8.

Listing 13.8   `traveler_comb.m` (available at
`http://physics.wm.edu/programming_with_MATLAB_book/./`
`ch_optimization/code/traveler_comb.m`)

```
function [best_route, shortest_distance]=traveler_comb(x,y
    );
% x - cities x coordinates
% y - cities y coordinates

% helper function
function dist=route_distance(route)
    dx=diff( x(route) );
    dy=diff( y(route) );
```

```
        dist = sum( sqrt(dx.^2 + dy.^2) );
end


% initialization
N=length(x); % number of cities
init_sequence=1:N;

p=init_sequence(2:N-1); % since we start at the 1st city
    and finish in the last
pold=p*0; % pold MUST not be equal to p

route=[1,p,N]; % any route is better than none
best_route=route;
shortest_distance=route_distance(route);
% show the initial route with the first and the last
    cities marked with 'x'
plot( x(1), y(1), 'x', x(N), y(N), 'x',  x(2:N-1), y(2:N
    -1), 'o', x(route), y(route), '-');

while ( any(pold ~=p) ) % as long as the new permutation
    is different from the old one
    % Notice the 'any' operator above.
    pold=p;
    p=permutation(pold);
    route=[1,p,N];
    dist=route_distance(route);
    if (dist < shortest_distance)
        shortest_distance=dist;
        best_route=route;
        % Uncomment the following lines to see the
            currently best route
        %plot( x(1), y(1), 'x', x(N), y(N), 'x',  x(2:N-1)
            , y(2:N-1), 'o', x(route), y(route), '-');
        %drawnow; % forces the figure update
    end
end
% plot all the cities and the best route
plot( x(1), y(1), 'x', x(N), y(N), 'x',  x(2:N-1), y(2:N
    -1), 'o', x(best_route), y(best_route), '-');

end
```

The author would like to attract the reader's attention to the helper function route_distance, which calculates the route distance, as the name suggests. Here, we piggyback on MATLAB's ability to generate an array with elements output in

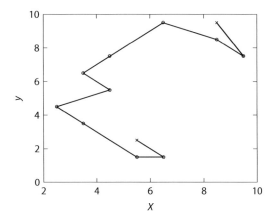

Figure 13.9   The shortest route connecting 12 cities. The first and the last cities to visit are marked with a cross, and the others are marked with a circle.

the order specified by the array of indexes (route in this case). The code also can plot the currently found best route as it executes.

Let's see how it copes with selection of the shortest route connecting 12 cities. We assign their coordinates rather randomly for the following test:

```
>> x = [8.5 3.5 9.5 4.5 2.5 3.5 6.5 5.5 4.5 8.5 6.5 5.5];
>> y = [9.5 3.5 7.5 7.5 4.5 6.5 1.5 1.5 5.5 8.5 9.5 2.5];
>> the_shortest_route = traveler_comb(x,y)
    the_shortest_route = [1 3 10 11 4 6 9 5 2 8 7 12]
```

The shortest route connecting the cities with the given coordinates is shown in Figure 13.9. The author's computer takes about 90 seconds to find this route.

13.5 Simulated Annealing Algorithm

We see that probing the full space permitted by combinatorics is not practical even for a seemingly small set of options. However, nature seems to handle the problem of energy minimization without any trouble. For example, if you think about a piece of metal, it has many atoms (the Avogadro number 6×10^{23} gives an order of magnitude estimate). Each of the atoms can be in many different states. So, the problem space must be humongous. Yet, if we slowly cool the metal, that is, anneal it, then the system will reach the minimum energy state.

In 1953, Metropolis and coworkers suggested an algorithm that can mimic the distribution of system states according to energies of the states and the overall temperature of the whole physical system (see [8]), that is, according to the Boltzmann energy distribution law. This law states that the probability of having energy E is given by

$$p(E) \sim \exp\left(-\frac{E - E_0}{kT}\right) \tag{13.20}$$

where:

E_0 is the energy of the lowest energy state

k is the Boltzmann constant

T is the temperature of the system.[*]

Recall that one of the names of the merit function is energy, which is very handy here. Note that if the temperature goes to zero, the probability of an energy state any higher than the global minimum drops to zero according to the above equation. Now, we have an idea for the general algorithm: evolve the system according to the Metropolis algorithm to mimic its physical behavior and simultaneously lower the temperature (anneal) to force the system into the lowest energy state. We spell out all the steps of this algorithm in the following box.

Simulated annealing or modified Metropolis algorithm

1. Set the temperature to a high value, so kT is larger than typical energy (merit) function fluctuation.

   - This requires some experiments if you do not know this a priori.

2. Assign a state $\vec{x}$ and calculate its energy $E(\vec{x})$.

3. Change, somehow, the old $\vec{x}$ to generate a new one, $\vec{x}_{new}$.

   - $\vec{x}_{new}$ should be somewhat **close or related** to the old optimal $\vec{x}$.

4. Calculate the energy at the new point $E_{new} = E(\vec{x})$.

5. If $E_{new} < E$, then $x = x_{new}$ and $E = E_{new}$.

   - That is, we move to the new point of the lower energy;

   otherwise, move to the new point with probability

$$p = \exp\left(-\frac{E_{new} - E}{kT}\right) \qquad (13.21)$$

6. Decrease the temperature a bit, that is, keep annealing.

7. Repeat from step 3 for a given number of cycles.

8. $\vec{x}$ will hold the local optimum solution.

You are probably wondering what the temperature of an optimization problem is. Well, it is just a parameter, that is, a number, which a physicist would insist on calling temperature because of its overall resemblance to the one in physics. So, do not worry—you do not need to put a thermometer inside of your computer.

---

[*] The derivation of this law is done in Statistical Mechanics and Thermodynamics courses.

In finite time (limited number of cycles), the algorithm is guaranteed to find only the local minimum.* But there is a theorem (see [6]) that states:

> The probability of finding the best solution goes to 1 if we run the algorithm for a longer and longer time with a slower and slower rate of cooling.

Unfortunately, this theorem is of no use, since it does not give a constrictive recipe of how long to run the algorithm. It is even suggested that it will need more cycles than the brute force combinatorial search. However, in practice, a good enough solution, that is, quite close to the global minimum, can be found in quite a short time with a quite small number of cycles.

A nice feature of the simulated annealing algorithm is that it is not limited to discrete space problems and can be used for problems accepting real values of $\vec{x}$ components. The algorithm also has the ability to climb away from a local minimum if it is given enough time.

To make an efficient (fast) implementation of this algorithm, we need to choose the optimal cooling rate* and the proper way to modify $\vec{x}$ so that the new value is not changing too much, that is, majority of the time we are in the vicinity of the optimal solution. It is quite challenging to make the right choices.

13.5.1 The backpack problem solution with the annealing algorithm

We will show the solution of the backpack problem with the simulated annealing algorithm in Listing 13.9. As we discussed Section 13.4.1, the main challenge is to find a good routine to generate a new candidate for the $\vec{x}_{new}$, which should be related to the previous best $\vec{x}$. We do not want to randomly sample arbitrary positions of the problem space.

Recall that $\vec{x}$ generally looks like $[0, 1, 1, 0, 1, \cdots, 0, 1, 1]$, so we should randomly toggle or mutate a small subset of the bits. We do it randomly,[†] so we do not need to keep track of flipped or not flipped positions. This is done by the change_x subfunction in Listing 13.9.

The rest is quite straightforward, as long as we remember that we are looking for the maximum value in the backpack. The Metropolis algorithm is designed for merit function minimization. So, we choose our merit function to be the negative value of all items in the backpack. Note that a random mutation could lead to a state with an overfilled backpack. So, we need to add a penalty for the case of the overfilled backpack. The positive number proportional to the overfilled portion is a good choice, which is a good way to send feedback to the minimization algorithm that such states are not welcomed.

---

* Actually, if the final temperature is not zero, the final x could be away from the minimum due to a finite probability of going to the higher energy state at step 5.
* If it is too fast, we will get stuck in a local minimum, and if it is too slow, we will waste a lot of central processing unit (CPU) cycles by probing around the global minimum.
† If you have a choice to make and you cannot reason which is better, then make a decision by a coin flip, that is, randomly. After all, any solution is better than none.

Words of wisdom
The penalty points that are usually given for incorrect homework assignment completion, should be called *negative feedback strength points*. They help students to see how far from the optimum they are. Also, control theory teaches us that the most effective feedback is negative feedback.

All of these points are taken care of in the backpack_merit subfunction shown in Listing 13.9. The rest is just bookkeeping and the straightforward realization of the simulated annealing algorithm. The code for the backpack problem implementing this method is shown in Listing 13.9.

Listing 13.9 backpack_metropolis.m (available at http://physics.wm.edu/ programming_with_MATLAB_book/./ch_optimization/code/ backpack_metropolis.m)

```
function [items_to_take, max_packed_value,
   max_packed_volume] = backpack_metropolis( backpack_size
   , volumes, values )
% Returns the list of items which fit in the  backpack and
   have the maximum total value.
% Solving the backpack problem with the simulated
   annealing (aka Metropolis) algorithm.
% backpack_size - the total volume of backpack
% volumes       - the vector of items volumes
% values        - the vector of items values

N=length(volumes); % number of items

function xnew=change_x(xold)
    % x is the state vector consisting of the take or no
       take flags
    % (i.e. 0/1 values) for each item
    % The new vector will be generated via random mutation
    % of every take or no take flag of the old one.
    flip_probability = 1./N; % in average 1 bit will be
       flipped
    bits_to_flip = (rand(1,N) < flip_probability );
    xnew=xold;
    xnew(bits_to_flip)=xor( xold(bits_to_flip) , 1 ); %
       xor operator flips the chosen flags
    if ( any( xnew ~= xold) )
            % at least 1 flag is flipped, so we are good to
               return
            return;
        else
```

```
            % none of the flags is flipped, so we try again
            xnew=change_x(xold); % recursive call to itself
        end
end

function [E,  items_value, items_volume] = backpack_merit(
    x, backpack_size, volumes, values)
    % Calculates the merit function for the backpack
        problem
    items_volume=sum(volumes .* x);
    items_value=sum(values  .* x);
    % The Metropolis algorithm is the minimization
        algorithm,
    % thus, we flip the packed items value (which we are
        maximizing)
    % to make the merit function to be minimization
        algorithm compatible.
    E= - items_value;

    % we should take care of the situations when the
        backpack is overfilled
    if ( (items_volume > backpack_size) )
        % Items do not fit and backpack, i.e. bad choice
            of the input vector 'x'.
        % We need to add a penalty.
        penalty=(items_volume-backpack_size); % overfill
            penalty
        % The penalty coefficient (mu) must be quite big,
        % but not too big or we will get stack in a local
            minimum.
        % Choosing this coefficient require a little
            tweaking and
        % depends on size of backpack, values and volumes
            vectors
        mu=100;
        E=E+mu*penalty;
    end
end


%% Initialization
% the current 'x' is the best one, since no other choices
    were checked.
xbest=zeros(1,N);
```

```matlab
[Ebest, max_packed_value, max_packed_volume]=
    backpack_merit(xbest, backpack_size, volumes, values);

Ncycles=10000; % number of annealing cycles
kT=max(values)*5; % should be large enough to permit even
    large and non optimal merit values
kTmin=min(values)/5; % should be smaller than the smallest
    step in energy
% we choose annealing coefficient by solving: kTmin=kT*
    annealing_coef^Ncycles
annealing_coef= power(kTmin/kT, 1/Ncycles); % the
    temperature lowering rate

best_energy_at_cycle=NaN(1,Ncycles); % this array is used
    for illustrations of the annealing

% the main annealing cycle
for c=1:Ncycles
    xnew=change_x(xbest);
    [Enew, items_value_new, items_volume_new] = ...
            backpack_merit(xnew, backpack_size, volumes,
                values);

    prob=rand(1,1);
    if ( (Enew < Ebest) || ( prob < exp(-(Enew-Ebest)/kT)
        ) )
        % Either this point has smaller energy
        %  and we go there without thinking
        % or
        %  according to the Metropolis algorithm
        %  there is the probability exp(-dE/kT) to move
            away from the current optimum
        xbest = xnew;
        Ebest = Enew;
        max_packed_value=items_value_new;
        max_packed_volume=items_volume_new;
    end
    % anneal or cool the temperature
    kT=annealing_coef*kT;

    best_energy_at_cycle(c)=Ebest; % keeping track of the
        current best energy value
end
plot(1:Ncycles, best_energy_at_cycle); % the annealing
    illustrating plot
```

```
xlabel('Cycle number');
ylabel('Energy');

% the Metropolis algorithm can return a non valid solution
  ,
% i.e. with combined volume larger than the volume of the
  backpack.
% For simplicity, no checks are done to prevent it.
indexes=1:N;
items_to_take=indexes(xbest==1);

end
```

At first, we will test our code with the same inputs as we did for the binary search algorithm in Section 13.5.1.

```
>> backpack_size=7;
>> volumes=[  2,  5,  1,  3, 3];
>> values =[ 10, 12, 23, 45, 4];
>> [items_to_take, max_packed_value] = ...
        backpack_metropolis( backpack_size, volumes,
           values)

  items_to_take = [1 3 4]
  max_packed_value = 78
```

As you can see, the result is exactly the same as in the case of the search over the full combinatorial space. This is not too surprising, since we did 10,000 cycles of probing (or annealing) for the problem with only 5 items, whose parameter space is $2^5 = 32$. Let's test it with the 20 items problem:

```
>> Vb=35;
>> val = [ 12 13 22 24 97 30 21 67 91 43 36 10 52 30 15 73 43 25 55 6];
>> vol = [ 20 27 34 23 4 22 32 2 30 34 34 24  8 23 18 30 14 4 27 22];
>> tic; [items, max_val, max_vol] = backpack_binary(Vb, vol, val); toc
   Elapsed time is 23.823041 seconds.
>>  items
  items = 5     8    13    17    18
>> max_val
  max_val = 284
>> tic; [items, max_val, max_vol] = backpack_metropolis(Vb,vol,val);toc
   Elapsed time is 0.515279 seconds.
>> items
   items = 5     8    13    17    18
>> max_val
  max_val = 284
```

As we can see, both algorithms produced the same result, that is, the list of items to choose and the maximum packed value of 284. Your answer might be slightly different when you run backpack_metropolis, since there is a small probability that on the last step the algorithm will end up with a less favorable energy state (i.e., away from optimal). The binary search takes more than 20 seconds, while the simulated annealing search takes only half a second. The best part is that, even for the larger problem with more items to choose from, it will still take only half a second. So, the small probability of getting sub-optimal, but still very good, results is a small price to pay for the drastic increase in speed.

You might have noticed that the backpack_metropolis function produces the plot of the merit or energy of the state used at the given cycle number of the annealing. This plot is very useful to help in judging whether the speed of annealing has been chosen properly. Let's have a look at Figure 13.10. These plots are generated for the problem with 20 items and exactly the same code of backpack_metropolis, with only one difference: the number of cycles Ncycles chosen was 100, 1,000, and 10,000. When we choose to do only 100 cycles, the algorithm quickly locks itself in a local minimum (as shown in the left insert) with energy somewhat higher than the lowest possible energy of −284. For the 1,000 cycles case (shown in the middle insert), the algorithm explores energy space and goes quite high in energy, but after about 200 cycles, it starts to search around the global minimum. In the

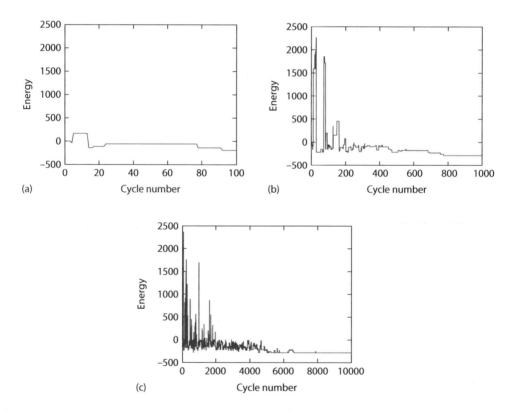

Figure 13.10   The current lowest energy state vs. the annealing cycle number for different total numbers of annealing cycles: 100 (a), 1,000 (c), and 10,000 (b).

last case of 10,000 cycles (shown in the right insert), the story is somewhat similar, except that we converge to global minimum only after about 6000 cycles.

So, we would say that 100 cycles are too few to cool the system sufficiently. The 10,000 cycles case seems to be cooling too slowly, since we spend a lot of cycles wandering around. However, the probability of ending up in the global minimum is the highest in this case. The 1,000 cycles case seems to be the best for this particular set of input parameters, since we find a good answer much faster than with 10,000 cycles (and the solution's energy seems to be the same), while a run with 100 cycles produces a quick, but sub-optimal, solution.

Generally, we would like the state energy to behave similarly to the middle and the right insert of Figure 13.10, where the energy oscillates at the beginning and then moves mostly downwards toward the minimum (global or maximum). Behavior like this is the sign of the proper choice of the annealing rate.

Another tricky part is the proper selection of the initial and final temperatures. Have a look at Figure 13.11. These plots are all produced with the same code as in Listing 13.9 but with only 100 overall cycles. In one case, we reduce both the initial and final temperatures by 1000 times, and in the other case, both temperatures

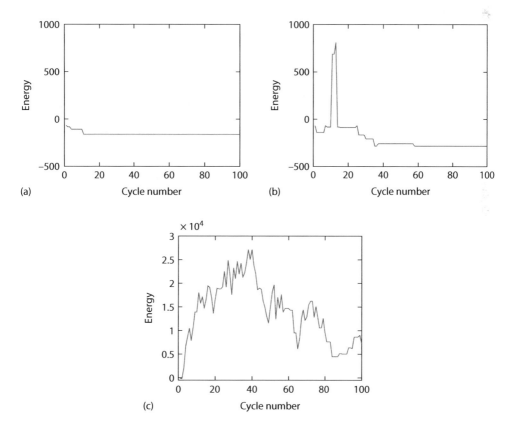

Figure 13.11 The current lowest energy state vs. the annealing cycle number for different annealing temperatures: in the left insert, the temperature's 1000 times smaller than in the middle, and in the right insert, the temperature is 1000 times larger. In all cases, the simulation is run for 100 cycles.

are 1000 times higher. For the case of the low temperatures (the left insert), the algorithm locks itself in the local minimum, which is higher than the global one. You can see this by the presence of only downward changes in the energy plot. For the higher temperature case (the right insert), the algorithm is continuously jumping up and down, since for high temperatures, upward motion is almost as likely as downward motion. Essentially, the temperature is never cold enough for the system to settle in any minimum. It is clear that the resulting solution for this case is the worst one: notice the scale for the energy and the fact that final energy is positive, that is, this is an illegal solution with an overfilled backpack. The middle insert, with the intermediate temperature settings, shows the behavior when the temperature parameters are better tuned: the energy climbs out of the local minimum around cycle 15 and then has **mostly** downhill dynamics with decreasing energy as we approach the end of the annealing.

13.6   Genetic Algorithm

The idea of the genetic algorithm is taken from nature, which is usually able to find the optimal solution via natural selection (see [5]). This algorithm has many modifications, but the main idea is shown in the following box.

Genetic algorithm

1. Generate a population (set of $\{\vec{x}\}$).
   - It is up to you to decide how large this set should be.
2. Find the fitness (i.e., merit) function for each member of the population.
3. Remove from the pool all but the most fit.
   - How many should stay is up to heuristic tweaks.
4. From the most fitted* (parents) breed a new population (children) to the size of the original population.
5. Repeat several times starting from step 2.
6. Select the fittest member of your population to be the final solution.

As usual, the trickiest part is to generate a new $\vec{x}$, that is, a child, from the older ones. Let's use the recipe provided by nature. We will refer to $\vec{x}$ as a chromosome or genome (hence the name of the algorithm).

---

* We are literally implementing "the survival of the fittest." Thus, the name for the merit or energy function is "fitness" for the genetic algorithm.

Generation of the children's genomes

1. Choose two parents randomly from the most fit set.

2. Crossover or recombine parents, chromosomes: take genes (i.e., $\vec{x}$ components) randomly from either of the parents and assign them to the new child chromosome.

3. Mutate (i.e., change) randomly some of the child's genes.

Some algorithm modifications allow parents to be in the new cycle of selection, while others eliminate them in the hope of moving away from a local minimum.

To find a good solution, you need a large population, since this lets you explore a larger parameter space. Think about the evolution strategies of microbes versus humans. However, this in turn leads to a longer computational time for every selection cycle. A nice feature of the genetic algorithm is that it suits the parallel computation paradigm: you can evaluate the fitness of each child on a different CPU and then compare their fitnesses.

As in any other optimum search algorithm except the full combinatorial search, the genetic algorithm is not guaranteed to find the global optimum in finite time.

13.7   Self-Study

General comments:

- Do not forget to run some test cases.

Problem 13.1
Prove (analytically) that the golden section algorithm R is still given by the same expression even if we need to choose $a' = x_1$ and $b' = b$.

Problem 13.2
Assume that the initial spacing between initial bracket points is h. Estimate (analytically) how many iterations it requires to narrow the bracket to the $10^{-9} \times h$ space.

Problem 13.3
Implement the golden section algorithm. Do not forget to check your code with simple test cases. Find where the function $E1(x) = x^2 - 100 * (1 - \exp(-x))$ has a minimum.

Problem 13.4
For the coin flipping game described in Section 12.2, find the optimal (maximizing your gain) betting fraction using the golden section algorithm and Monte Carlo simulation. Feel free to reuse the complimentary codes provided.

Note: you need a lot of game runs to have reasonably small uncertainty for the merit function evaluations. I would suggest averaging at least 1000 runs with the length of 100 coin flips each.

Problem 13.5

Find the point where the function

$$F(x, y, z, w, u) = (x - 3)^2 + (y - 1)^4 + (u - z)^2 + (u - 2 * w)^2 + (u - 6)^2 + 12$$

has a minimum. What is the value of $F(x, y, z, w, u)$ at this point?

Problem 13.6

Modify the provided traveling salesman combinatorial algorithm to solve a slightly different problem. You are looking for the shortest route that goes through all cities while it starts and ends in the same city (the first one), that is, we need a close loop route.

Coordinates of the cities are provided in the 'cities_for_combinatorial_search.dat'.[*] file: the first column of the data file corresponds to the x coordinate and the second one to the y coordinate. The coordinates of the city where the route begins and ends are in the first row.

Provide your answers to the following questions:

- What is the sequence of all cities in the shortest route?

- What is the total length of the best route?

- Provide the plot with the visible cities' locations and the shortest route.

Problem 13.7

Implement the Metropolis algorithm to solve problem 13.6. A good way to obtain a new test route is to randomly swap two cities along the route. You need to choose the number of cycles and initial and final temperature (kT). Provide the reasons for your choices.

As a test, compare this algorithm's solution with the combinatorial solution.

Now, load the cities, coordinates from the 'cities_for_metropolis_search.dat'.[†] file. Find the shortest route for this set of cities.

- What is the sequence of all cities in the shortest route?

- What is the total length of the best route?

- Provide the plot with the visible cities' locations and the shortest route.

---

[*] The file is available at http://physics.wm.edu/programming_with_MATLAB_book/./ ch_optimization/data/cities_for_combinatorial_search.dat
[†] The file is available at http://physics.wm.edu/programming_with_MATLAB_book/./ ch_optimization/data/cities_for_metropolis_search.dat

CHAPTER 14

Ordinary Differential Equations

This chapter discusses methods to solve ordinary differential equations. It explains the classic Euler and Runge–Kutta methods as well as MATLAB's built-in commands. We show how they are used with examples of free fall and motions with air drag.

14.1 Introduction to Ordinary Differential Equation

In mathematics, an ordinary differential equation (ODE) is an equation that contains functions of only one variable and its derivatives.

> **An ordinary differential equation of order n has the following form:**
>
> $$y^{(n)} = f(x, y, y', y'', \cdots, y^{(n-1)}) \tag{14.1}$$
>
> where:
>
> x is the independent variable
>
> $y^{(i)} = \frac{d^i y}{dx^i}$ is the ith derivative of $y(x)$
>
> f is the force term

> **Example**
>
> Arguably, the most famous ODE of the second order is Newton's second law connecting the acceleration of the body (a) to the force (F) acting on it:
>
> $$a(t) = \frac{F}{m}$$
>
> where
>
> m is the mass of the body
>
> t is time
>
> For simplicity, we talk here only about the y component of the position. Recall that acceleration is the second derivative of the position, that is, $a(t) = y''(t)$. The force function F might depend on time, the body position, and its velocity (i.e., the first derivative of position y'), so F should be written as $F(t, y, y')$. So, we rewrite Newton's second law as the second-order ODE:
>
> $$y'' = \frac{F(t, y, y')}{m} = f(t, y, y') \tag{14.2}$$
>
> As you can see, time serves as independent variable. We can obtain the canonical Equation 14.1 by simply relabeling $t \rightarrow x$.

Any nth-order ODE (such as Equation 14.1) can be transformed to a system of first-order ODEs.

Transformation of nth-order ODE to a system of first-order ODEs

We define the following variables:

$$y_1 = y, y_2 = y', y_3 = y'', \cdots, y_n = y^{(n-1)} \tag{14.3}$$

Then, we can write the following system of equations:

$$\begin{pmatrix} y_1' \\ y_2' \\ y_3' \\ \vdots \\ y_{n-1}' \\ y_n' \end{pmatrix} = \begin{pmatrix} f_1 \\ f_2 \\ f_3 \\ \vdots \\ f_{n-1} \\ f_n \end{pmatrix} = \begin{pmatrix} y_2 \\ y_3 \\ y_4 \\ \vdots \\ y_n \\ f(x, y_1, y_2, y_3, \cdots y_n) \end{pmatrix} \tag{14.4}$$

We can rewrite Equation 14.4 in a much more compact vector form.

The canonical form of the ODE system

$$\vec{y}' = \vec{f}(x, \vec{y}) \tag{14.5}$$

Example

Let's convert Newton's second law (Equation 14.2) to a system of first-order ODEs. The acceleration of a body is the first derivative of velocity with respect to time and is equal to the force divided by mass:

$$\frac{dv}{dt} = v'(t) = a(t) = \frac{F}{m}$$

Also, we recall that velocity itself is the derivative of the position with respect to time.

$$\frac{dy}{dt} = y'(t) = v(t)$$

Combining these, we rewrite Equation 14.2 as

$$\begin{pmatrix} y' \\ v' \end{pmatrix} = \begin{pmatrix} v \\ f(t, y, v) \end{pmatrix} \tag{14.6}$$

We do the following variable relabeling: $t \rightarrow x$, $y \rightarrow y_1$, and $v \rightarrow y_2$, and rewrite our equation in the canonical form resembling Equation 14.4:

$$\begin{pmatrix} y_1' \\ y_2' \end{pmatrix} = \begin{pmatrix} y_2 \\ f(x, y_1, y_2) \end{pmatrix} \tag{14.7}$$

14.2 Boundary Conditions

The system of n ODEs requires n constraints to be fully defined. This is done by providing the *boundary conditions*. There are several alternative ways to do it. The most intuitive way is by specifying the full set of $\vec{y}$ components at some starting position x_0, that is, $\vec{y}(x_0) = \vec{y}_0$. This is called the *initial value problem*.

The alternative way is to specify some components of $\vec{y}$ at the starting value x_0 and the rest at the final value x_f. The problem specified in this way is called the *two-point boundary value problem*.

In this chapter, we will consider only the initial value problem and its solutions.

The initial value problem boundary conditions

We need to specify all components of the $\vec{y}$ at the initial position x_0.

$$\begin{pmatrix} y_1(x_0) \\ y_2(x_0) \\ y_3(x_0) \\ \vdots \\ y_n(x_0) \end{pmatrix} = \begin{pmatrix} y_{1_0} \\ y_{2_0} \\ y_{3_0} \\ \vdots \\ y_{n_0} \end{pmatrix} = \begin{pmatrix} y_0 \\ y_0' \\ y_0'' \\ \vdots \\ y_0^{(n-1)} \end{pmatrix}$$

For the Newton's second law example, which we considered in the previous section, the boundary condition requires us to specify the initial position and velocity of the object in addition to Equation 14.7. Then, the system is fully defined and has only one possible solution.

14.3 Numerical Method to Solve ODEs

14.3.1 Euler's method

Let's consider the simplest case: a first-order ODE (notice the lack of the vector notation)

$$y' = f(x, y)$$

There is an exact way to write the solution:

$$y(x_f) = y(x_0) + \int_{x_0}^{x_f} f(x, y) dx$$

The problem with this formula is that the $f(x, y)$ depends on the y itself. However, on a small enough interval $[x, x + h]$, we can assume that $f(x, y)$ does not change, that is, it is constant. In this case, we can use the familiar rectangles integration formula (see Section 9.2):

$$y(x + h) = y(x) + f(x, y)h$$

When applied to an ODE, this process is called *Euler's method*.

We need to split our $[x_0, x_f]$ interval into a bunch of steps of the size h and leap frog from the x_0 to the $x_0 + h$, then to the $x_0 + 2h$, and so on.

Now, we can make an easy transformation to the vector case (i.e., the nth-order ODE):

Euler's method (error $\mathcal{O}(h^2)$)

$$\vec{y}(x+h) = \vec{y}(x) + \vec{f}(x,y)h$$

The MATLAB implementation of Euler's method is shown in Listing 14.1.

Listing 14.1 `odeeuler.m` (available at `http://physics.wm.edu/`
`programming_with_MATLAB_book/./ch_ode/code/odeeuler.m`)

```
function [x,y]= odeeuler(fvec, xspan, y0, N)
    %% Solves a system of ordinary differential equations
        with the Euler method
    % x - column vector of x positions
    % y - solution array values of y, each row corresponds
        to particular row of x.
    %       each column corresponds, to a given derivative
        of y,
    %       including y(:,1) with no derivative
    % fvec   - handle to a function f(x,y) returning forces
        column vector
    % xspan - vector with initial and final x coordinates
        i.e. [x0, xf]
    % y0      - initial conditions for y, should be row
        vector
    % N       - number of points in the x column (N>=2),
    %           i.e. we do N-1 steps during the calculation

    x0=xspan(1);            % start position
    xf=xspan(2);            % final position

    h=(xf-x0)/(N-1);        % step size
    x=linspace(x0,xf,N);    % values of x where y will be
        evaluated

    odeorder=length(y0);
    y=zeros(N,odeorder);    % initialization
    x(1)=x0; y(1,:)=y0;     % initial conditions
```

```
   for i=2:N % number of steps is less by 1 then number
       of points since we know x0,y0
       xprev=x(i-1);
       yprev=y(i-1,:);
       % Matlab somehow always send column vector for 'y'
           to the forces calculation code
       % transposing yprev to make this method compatible
           with Matlab.
       % Note the dot in .' this avoid complex conjugate
           transpose
       f=fvec(xprev, yprev.');
       % we receive f as a column vector, thus, we need
           to transpose again
       f=f.';
       ynext=yprev+f*h;   % vector of new values of y: y(x
           +h)=y(x)+f*h
       y(i,:)=ynext;
   end
end
```

Similarly to the rectangle integration method, which is inferior in comparison to more advanced methods (e.g., the trapezoidal and Simpson's), the Euler method is less precise for a given h. There are better algorithms, which we will now discuss.

14.3.2 The second-order Runge–Kutta method (RK2)

Using multivariable calculus and the Taylor expansion, as shown for example in [1], we can write

$$\vec{y}(x_{i+1}) = \vec{y}(x_i + h) =$$
$$= \vec{y}(x_i) + C_0\vec{f}(x_i, \vec{y}_i)h + C_1\vec{f}(x_i + ph, \vec{y}_i + qh\vec{f}(x_i, \vec{y}_i))h + \mathcal{O}(h^3)$$

where C_0, C_1, p, and q are some constants satisfying the following set of constraints:

$$C_0 + C_1 = 1 \tag{14.8}$$
$$C_1 p = 1/2 \tag{14.9}$$
$$C_1 q = 1/2 \tag{14.10}$$

It is clear that the system is under-constrained, since we have only three equations for four constants. There are a lot of possible choices of parameters C_0, C_1, p, and q. One choice has no advantage over another.

But there is one "intuitive" choice: $C_0 = 0$, $C_1 = 1$, $p = 1/2$, and $q = 1/2$. It provides the following recipe for how to find $\vec{y}$ at the next position after the step h.

Modified Euler's method or midpoint method (error $\mathcal{O}(h^3)$)

$$\vec{k}_1 = h\vec{f}(x_i, \vec{y}_i)$$

$$\vec{k}_2 = h\vec{f}(x_i + \frac{h}{2}, \vec{y}_i + \frac{1}{2}\vec{k}_1)$$

$$\vec{y}(x_i + h) = \vec{y}_i + \vec{k}_2$$

As the name suggests, we calculate what $\vec{y}(x + h)$ could be with the Euler-like method by calculating $\vec{k}_1$, but then we do only a half step in that direction and calculate the updated force vector in the midpoint. Finally, we use this force vector to find the improved value of $\vec{y}$ at $x + h$.

14.3.3 The fourth-order Runge-Kutta method (RK4)

A higher-order expansion of the $\vec{y}(x + h)$ also allows multiple choices of possible expansion coefficients (see [1]). One of the "canonical" choices (see [9]) is spelled out in

---

The fourth-order Runge–Kutta method with truncation error $\mathcal{O}(h^5)$

$$\vec{k}_1 = h\vec{f}(x_i, \vec{y}_i)$$

$$\vec{k}_2 = h\vec{f}(x_i + \frac{h}{2}, \vec{y}_i + \frac{1}{2}\vec{k}_1)$$

$$\vec{k}_3 = h\vec{f}(x_i + \frac{h}{2}, \vec{y}_i + \frac{1}{2}\vec{k}_2)$$

$$\vec{k}_4 = h\vec{f}(x_i + h, \vec{y}_i + \vec{k}_3)$$

$$\vec{y}(x_i + h) = \vec{y}_i + \frac{1}{6}(\vec{k}_1 + 2\vec{k}_2 + 2\vec{k}_3 + \vec{k}_4)$$

14.3.4 Other numerical solvers

We have by no means covered all methods of solving ODEs. So far, we have only talked about fixed step *explicit* methods. When the force term is changing slowly, it is reasonable to increase the step size h, or decrease it when the force term is quickly varying at the given interval. This leads to a slew of adaptive methods.

There are also *implicit* methods, in which one solves for $\vec{y}(x_i + h)$ satisfying the following equation:

$$\vec{y}(x_i) = \vec{y}(x_i + h) - f(x, \vec{y}(x_i + h))h \tag{14.11}$$

Such implicit methods are more robust, but they are computationally more demanding. Several other ODE solving algorithms are covered, for example, in [1, 9].

14.4   Stiff ODEs and Stability Issues of the Numerical Solution

Let's have a look at the first-order ODE

$$y' = 3y - 4e^{-x} \tag{14.12}$$

It has the following analytical solution:

$$y = Ce^{3x} + e^{-x} \tag{14.13}$$

where C is a constant.

If the initial condition is $y(0) = 1$, then the solution is

$$y(x) = e^{-x}$$

The `ode_unstable_example.m` script (shown in Listing 14.2) calculates and plots the numerical and the analytical solutions.

Listing 14.2   `ode_unstable_example.m` (available at `http://physics.wm.edu/ programming_with_MATLAB_book/./ch_ode/code/ode_unstable_example.m`)

```
%% we are solving y'=3*y-4*exp(-x) with y(0)=1
y0=[1]; %y(0)=1
xspan=[0,2];

fvec=@(x,y) 3*y(1)-4*exp(-x);
% the fvec is scalar, there is no need to transpose it to
   make a column vector

Npoints=100;
[x,y] = odeeuler(fvec, xspan, y0, Npoints);

% general analytical solution is
% y(x)= C*epx(3*x)+exp(-x), where C is some constant
% from y(0)=1  follows C=0
yanalytical=exp(-x);
```

```
plot(x, y(:,1), '-', x, yanalytical, 'r.-');
set(gca,'fontsize',24);
legend('numerical','analytical');
xlabel('x');
ylabel('y');
title('y vs. x');
```

As we can see in Figure 14.1, the numerical solution diverges from the analytical one. Initially, we might think that this is due to a large step, h, or the use of the inferior Euler method. However, even if we decrease h (by increasing Npoints) or change the ODE solving algorithm, the growing discrepancy from the analytical solution will show up.

This discrepancy is a result of accumulated round-off errors (see Section 1.5). From a computer's point of view, due to accumulated errors at some point, the numerically calculated $y(x)$ deviates from the analytical solution. This is equivalent to saying that we follow the track where initial condition is $y(0) = 1 + \delta$. The δ is small, but it forces $C \neq 0$; thus, the numerical solution picks up the diverging $\exp(3x)$ term from Equation 14.13. We might think that the decrease of h should help, at least, this clearly push the deviation point to the right. This idea forces us to pick smaller and smaller h (thus, increasing the calculation time) for an otherwise seemingly smooth evolution of y and its derivative. These kinds of equations are called *stiff*. Note that due to the round-off errors, we cannot decrease h indefinitely. The implicit algorithms (which are only briefly mentioned in Section 14.3.4) usually are more stable in such settings.

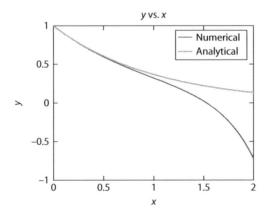

Figure 14.1   Comparison of numerical and analytical solutions of Equation 14.12.

Words of wisdom
Do not trust the numerical solutions (regardless of the method) without proper consideration.

14.5 MATLAB's Built-In ODE Solvers

Have a look in the help files for ODEs. In particular, pay attention to

- `ode45` uses adaptive explicit fourth-order Runge–Kutta method (good default method).

- `ode23` uses adaptive explicit second-order Runge–Kutta method.

- `ode113` is suitable for "stiff" problems.

"Adaptive" means that you do not need to choose the step size h. The algorithm does it by itself. However, remember the rule about not trusting a computer's choice.

Run the built-in `odeexamples` command to see some of the demos for ODE solvers.

14.6 ODE Examples

In this section, we will cover several examples of physical systems involving ODEs. With any ODE, the main challenge is to transform a compact human notation to the canonical ODE form (see Equation 14.5), for which we have multiple numerical methods to produce the solution.

14.6.1 Free fall example

Let's consider a body that falls in Earth's gravitational field only in the vertical direction (y). For simplicity, we assume that there is no air resistance and the only force acting on the body is due to the gravitational pull of Earth. We also assume that everything happens at sea level, so the gravitational force is height independent. In this case, we can rewrite Newton's second law as

$$y'' = F_g/m = -g \tag{14.14}$$

where:

y is the vertical coordinate of the body
F_g is the force due to gravity
m is the mass of the body
$g = 9.8 \, \text{m/s}^2$ is the constant (under our assumptions) acceleration due to gravity

The gravitational force is directed opposite to the y coordinate axis, which points up; thus, we put the negative sign in front of g. The y'' is the second derivative with respect to time (the independent variable in this problem).

Equation 14.14 is a second-order ODE, which we need to convert to a system of two first-order ODEs. We note that the velocity component (v) along the y-axis is equal to the first derivative of the y position. The derivative of the velocity v is the acceleration y''. So, we can rewrite the second-order ODE as

$$\begin{pmatrix} y' \\ v' \end{pmatrix} = \begin{pmatrix} v \\ -g \end{pmatrix} \tag{14.15}$$

Finally, we transform this system to the canonical form of Equation 14.5 by the following relabeling: $t \rightarrow x$, $y \rightarrow y_1$, and $v \rightarrow y_2$:

$$\begin{pmatrix} y_1' \\ y_2' \end{pmatrix} = \begin{pmatrix} y_2 \\ -g \end{pmatrix} \tag{14.16}$$

To use an ODE numerical solver, we need to program the ODE force term calculation function, which is in charge of the right-hand side of this system of equations. This is done in Listing 14.3.

Listing 14.3 `free_fall_forces.m` (available at http://physics.wm.edu/ programming_with_MATLAB_book/./ch_ode/code/free_fall_forces.m)

```
function fvec=free_fall_forces(x,y)
        % free fall forces example
        % notice that physical meaning of the  independent
            variable 'x' is  time
        % we are solving y''(x)=-g, so the transformation
            to the canonical form is
        % y1=y; y2=y'
        % f=(y2,-g);

        g=9.8; % magnitude of the acceleration due to the
            free fall in m/s^2

        fvec(1)=y(2);
        fvec(2)=-g;
        % if we want to be compatible with Matlab solvers,
            fvec should be a column
        fvec=fvec.'; % Note the dot in .' This avoids
            complex conjugate transpose
end
```

Now, we are ready to numerically solve the ODEs. We do this with the algorithm implemented in `odeeuler.m` shown in Listing 14.1. MATLAB's built-ins are also perfectly suitable for this task, but we need to omit the number of points in this case, since the step size h is chosen by the algorithm. See how it is done in the code in Listing 14.4.

Listing 14.4 `ode_free_fall_example.m` (available at
`http://physics.wm.edu/programming_with_MATLAB_book/./ch_ode/code/`
`ode_free_fall_example.m`)

```
%% we are solving y''=-g,  i.e  free fall motion

% Initial conditions
y0=[500,15];  % we start from the height of 500 m  and our
    initial velocity is 15 m/s

% independent variable 'x' has the meaning of time in our
    case
timespan=[0,13]; % free fall for duration of  13 seconds


Npoints=20;

%% Solve the ODE
[time,y] = odeeuler(@free_fall_forces, timespan, y0,
    Npoints);
% We can use MATLAB's built-ins, for example ode45.
% In this case, we should  omit Npoints. See the line
    below.
%[time,y] = ode45(@free_fall_forces, timespan, y0);


%% Calculating the analytical solution
g=9.8;
yanalytical=y0(1) + y0(2)*time - g/2*time.^2;
vanalytical=y0(2) - g*time;

%% Plot the results
subplot(2,1,1);
plot(time, y(:,1), '-', time, yanalytical, 'r-');
set(gca,'fontsize',20);
legend('numerical','analytical');
xlabel('Time, S');
ylabel('y-position, m');
title('Position vs. time');
grid on;
```

```
subplot(2,1,2);
plot(time, y(:,2), '-', time, vanalytical, 'r-');
set(gca,'fontsize',20);
legend('numerical','analytical');
xlabel('Time, S');
ylabel('y-velocity, m/s');
title('Velocity vs. time');
grid on;
```

For such a simple problem, there is an exact analytical solution

$$\begin{cases} y(t) = y_0 + v_0 t - gt^2/2 \\ v(t) = v_0 - gt \end{cases} \tag{14.17}$$

The code in Listing 14.4 calculates and plots both the numerical and the analytical solutions to compare them against each other. The results are shown in Figure 14.2. As we can see, both solutions are almost on top of each other, that is, they are almost the same. Try to increase the number of points (Npoints), that is, decrease the step size h, to see how the numerical solution converges to the true analytical solution.

14.6.2 Motion with the air drag

The previous example is very simple. Let's solve a much more elaborate problem: the motion of a projectile influenced by air drag. We will consider two-dimensional motion in the x-y plane near the Earth's surface, so acceleration due to gravity

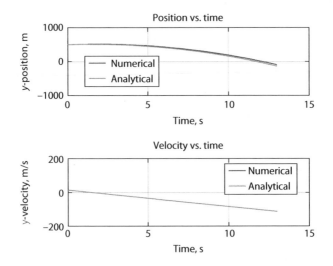

Figure 14.2   The free fall problem analytical and numerical solutions.

(g) is constant. This time, we have to take into account the air drag force $(\vec{F}_d)$, which is directed opposite to the body's velocity and proportional to the velocity $(\vec{v})$ squared (note the $v\vec{v}$ term in eq. (14.18)):

$$\vec{F}_d = -\frac{1}{2}\rho C_d A v\vec{v} \qquad (14.18)$$

here:

C_d is the drag coefficient, which depends on the projectile shape
A is the cross-sectional area of the projectile
ρ is the density of the air

For simplicity, we will assume that the air density is constant within the range of reachable positions.

Newton's second equation in this case can be written as

$$m\vec{r}\,'' = \vec{F}_g + \vec{F}_d \qquad (14.19)$$

where:

$\vec{r}$ is the radius vector tracking the position of the projectile
$\vec{F}_g = m\vec{g}$ is the force of the gravitational pull on the projectile with the mass m

This equation is second-order ODE. We transform it into a system of first-order ODEs similarly to the previous example:

$$\begin{pmatrix} \vec{r}\,' \\ \vec{v}\,' \end{pmatrix} = \begin{pmatrix} \vec{v} \\ \vec{F}_g/m + \vec{F}_d/m \end{pmatrix} \qquad (14.20)$$

We should pay attention to the vector form of these equations, which reminds us that each term has x and y components. We spell it out in the following equation:

$$\begin{pmatrix} x' \\ v_x' \\ y' \\ v_y' \end{pmatrix} = \begin{pmatrix} v_x \\ F_{g_x}/m + F_{d_x}/m \\ v_y \\ F_{g_y}/m + F_{d_y}/m \end{pmatrix} \qquad (14.21)$$

We can simplify this by noticing that $F_{g_x} = 0$, since the gravity is directed vertically. Also, we note that $F_{d_x} = -F_d v_x/v$ and $F_{d_y} = -F_d v_y/v$, where the magnitude of the air drag force is $F_d = C_d A v^2/2$. The simplified equation looks like

$$\begin{pmatrix} x' \\ v_x' \\ y' \\ v_y' \end{pmatrix} = \begin{pmatrix} v_x \\ -F_d v_x/(vm) \\ v_y \\ -g - F_d v_y/(vm) \end{pmatrix} \qquad (14.22)$$

Finally, we bring it to the canonical form with the following relabeling: $x \to y_1$, $v_x \to y_2$, $y \to y_3$, $v_y \to y_4$, and $t \to x$.

The key to success is to adhere to this transformation and alternate between it and the human (physical) notation during problem implementation. See how it is done in the code in Listing 14.5.

Listing 14.5 `ode_projectile_with_air_drag_model.m` (available at `http://` `physics.wm.edu/programming_with_MATLAB_book/./ch_ode/code/` `ode_projectile_with_air_drag_model.m`)

```
function [t, x, y, vx, vy] =
  ode_projectile_with_air_drag_model()
    %% Solves the equation of motions for a projectile
       with air drag included
    % r''= F = Fg+ Fd
    % where
    % r is the radius vector, Fg is the gravity pull force
      , and Fd is the air drag force.
    % The above equation can be decomposed to x and y
      projections
    % x'' = Fd_x
    % y'' = -g + Fd_y
    % Fd =  1/2 * rho * v^2 * Cd * A is the drag force
      magnitude
    % where v is speed.
    % The drag force directed against the velocity vector
    % Fd_x= - Fd * v_x/v  ; % vx/v takes care of the
      proper sign of the drag projection
    % Fd_y= - Fd * v_y/v  ; % vy/v takes care of the
      proper sign of the drag projection
    % where vx and vy are the velocity projections

    % at the first look it does not look like ODE but
      since x and y depends only on t
    % it is actually a system of ODEs


    % transform system to the canonical form
    % x   -> y1
    % vx -> y2
    % y   -> y3
    % vy -> y4
    % t   -> x
    %
    % f1 -> y2
    % f2 -> Fd_x
    % f3 -> y4
    % f4 -> -g + Fd_y
```

```
% some constants
rho=1.2;  % the density of air kg/m^3
Cd=.5;    % an arbitrary choice of the drag
   coefficient
m=0.01;   % the mass of the projectile  in kg
g=9.8;    % the acceleration due to gravity
A=.25e-4; % the area of the projectile in m^2, a
   typical bullet is 5mm x 5mm

function fvec = projectile_forces(x,y)
    % it is crucial to move from the ODE notation to
       the human notation
    vx=y(2);
    vy=y(4);
    v=sqrt(vx^2+vy^2); % the speed value

    Fd=1/2 * rho * v^2 * Cd * A;

    fvec(1) = y(2);
    fvec(2) = -Fd*vx/v/m;
    fvec(3) = y(4);
    fvec(4) = -g -Fd*vy/v/m;

    % To make matlab happy we need to return a column
       vector.
    % So, we transpose (note the dot in .')
    fvec=fvec.';
end

%% Problem parameters setup:
%  We will set initial conditions similar to a bullet
   fired from
%  a rifle at 45 degree to the horizon.
tspan=[0, 80];       % time interval of interest
theta=pi/4;          % the shooting angle above the
   horizon
v0 = 800;            % the initial projectile speed in
   m/s
y0(1)=0;             % the initial x position
y0(2)=v0*cos(theta); % the initial vx velocity
   projection
y0(3)=0;             % the initial y position
y0(4)=v0*sin(theta); % the initial vy velocity
   projection
```

```matlab
% We are using matlab solver
[t,ysol] = ode45(@projectile_forces, tspan, y0);

% Assigning the human readable variable names
x =  ysol(:,1);
vx = ysol(:,2);
y =  ysol(:,3);
vy = ysol(:,4);
v=sqrt(vx.^2+vy.^2); % speed

% The analytical drag-free motion solution.
% We should not be surprised by the projectile
   deviation from this trajectory
x_analytical = y0(1) + y0(2)*t;
y_analytical = y0(3) + y0(4)*t -g/2*t.^2;
v_analytical= sqrt(y0(2).^2 + (y0(4) - g*t).^2); %
   speed

ax(1)=subplot(2,1,1);
plot(x,y, 'r-', x_analytical, y_analytical, 'b-');
set(gca,'fontsize',14);
xlabel('Position x component, m');
ylabel('Position y component, m');
title ('Trajectory');
legend('with drag', 'no drag', 'Location','SouthEast')
   ;

ax(2)=subplot(2,1,2);
plot(x,v, 'r-', x_analytical, v_analytical, 'b-');
set(gca,'fontsize',14);
xlabel('Position x component, m');
ylabel('Speed');
title ('Speed vs. the x position component');
legend('with drag', 'no drag', 'Location','SouthEast')
   ;

linkaxes(ax,'x'); % very handy for related subplots
end
```

The code shows two trajectories of the same projectile: one for the case when we take into account the air drag and the other without it (see Figure 14.3). In the latter case, we provide the analytical solution as well as the numerical solution in a similar manner to the previous example. However, once we account for drag, we cannot easily determine the analytical solution and must solve numerically.

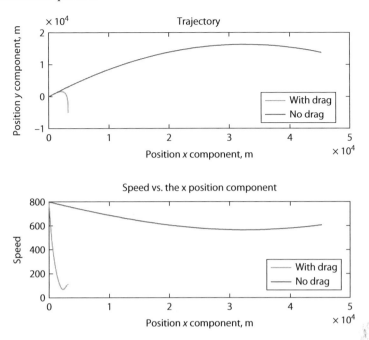

Figure 14.3   The projectile trajectory and its speed vs. x position calculated without and with the air drag force taken into account.

Finally, we justified the use of numerical methods.* The downside to solving the problem numerically is that we cannot easily check the result of numerical calculations in the same way that we could with the analytical solution. However, we still can do some checks, as we know that the drag slows down the projectile, so it should travel a shorter distance. Indeed, as we can see in Figure 14.3, the bullet travels only about 3 km when influenced by the drag, while drag-free bullets can fly further than 40 km.† Have a look at the trajectory of the bullet with the drag. Closer to the end, it drops straight down. This is because the drag reduces the x component of the velocity to zero, but gravity is still affecting the bullet, so it picks up the non-zero vertical component value. If we look at the speed plot, it is reasonable to expect that speed will decrease as the bullet travels. Why, then, does the speed increase at the end of the trajectory? We can see that this happens as the bullet reaches the highest point, so the potential energy converts to kinetic after this point, and the speed grows. The same effect is observed in the plot for the drag-free motion. The above sanity checks allow us to conclude that the numerical calculation seems to be reasonable.

---

* You might have noticed that the author provided an analytical solution, whenever it is possible, to be used as the test case for the numerical solution.

† The air not only lets us breathe but also allows us to live much closer to the firing ranges.

14.7　Self-Study

Problem 14.1

Here is a model for a more realistic pendulum. Numerically solve (using the built-in ode45 solver) the following physical problem of a pendulum's motions:

$$\theta''(t) = -\frac{g}{L}\sin(\theta)$$

where:

g　is acceleration due to gravity (g=9.8 m/s²)

$L = 1$　is the length of the pendulum

θ　is the angular deviation of the pendulum from the vertical

　　Assuming that the initial angular velocity (β) is zero, that is, $\beta(0) = \theta'(0) = 0$, solve this problem (i.e., plot $\theta(t)$ and $\beta(t)$) for two values of the initial deflection $\theta(0) = \pi/10$ and $\theta(0) = \pi/3$. Both solutions must be presented on the same plot. Make the final time large enough to include at least 10 periods. Show that the period of the pendulum depends on the initial deflection. Does it takes longer to make one swing with a larger or smaller initial deflection?

Problem 14.2

Implement the fourth-order Runge–Kutta method (RK4) according to the recipe outlined in Section 14.3.3. It should be input compatible to the home-made Euler's implementation Listing 14.1. Compare the solution of this problem with your own RK4 implementation to the built-in ode45 solver.

CHAPTER 15

Discrete Fourier Transform

This chapter discusses Fourier transform theory for continuous and discrete functions as well as Fourier transform applications. The chapter shows MATLAB's built-in methods for executing forward and inverse Fourier transforms.

We usually think about processes around us as functions of time. However, it is often useful to think about them as functions of frequencies. We naturally do this without giving it a second thought. For example, when we listen to someone's speech, we distinguish one person from another by the pitch, that is, dominating frequencies, of the voice. Similarly, our eyes do a similar time-to-frequency transformation, as we can distinguish different light colors, and the colors themselves are directly connected to the frequency of light. When we tune a radio, we select a particular subset of frequencies to listen in the time varying signal of electromagnetic waves. Even when we talk about salary, it is often enough to know that it will come every 2 weeks or a month, that is, we are concerned with the period and, thus, frequency of the wages.

The transformations from the time domain to the frequency domain and back are called the forward* Fourier and inverse Fourier transforms respectively. The transformation is general and can broadly be applied to any variable, not just a time variable. For example, we can do the transformation from spatial coordinates to the spatial coordinates' frequencies, which is the base of the JPEG image compression algorithm.

The Fourier transform provides the basis for many filtering and compression algorithms. It is also an indispensable tool for analyzing noisy data. Many differential equations can be solved much easier in the periodic oscillation basis. Additionally, the calculation of convolution integrals of two functions is very fast and simple when it is done in the frequency domain.

15.1 Fourier Series

It is natural to think that a periodic function, such as the example shown in Figure 15.1, can be constructed as the sum of other periodic functions. In Fourier series, we do it in the basis of sines and cosines, which are clearly periodic functions.

A more mathematically solid definition of a function that can be transformed is: any periodic single value function $y(t)$ with a finite number of discontinuities and for which $\int_0^T |f(t)|dt$ is finite can be presented as

---

* The word "forward" is often omitted.

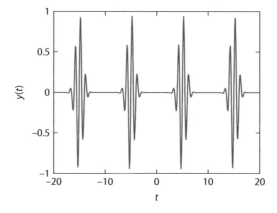

Figure 15.1   Example of a periodic function with the period of 10.

Fourier series

$$y(t) = \frac{a_0}{2} + \sum_{n=1}^{\infty} (a_n \cos(n\omega_1 t) + b_n \sin(n\omega_1 t)) \qquad (15.1)$$

where:

T is the period, that is, $y(t) = y(t + T)$

$\omega_1 = 2\pi/T$ is the fundamental angular frequency

constant coefficients a_n and b_n can be calculated according to the following formula:

$$\begin{pmatrix} a_n \\ b_n \end{pmatrix} = \frac{2}{T} \int_0^T \begin{pmatrix} \cos(n\omega_1 t) \\ \sin(n\omega_1 t) \end{pmatrix} y(t) dt \qquad (15.2)$$

At a discontinuity, the series approaches the midpoint, that is:

$$y(t) = \lim_{\delta \to 0} \frac{y(t - \delta) + y(t + \delta)}{2} \qquad (15.3)$$

Note that for any integer n, $\sin(n2\pi/Tt)$ and $\cos(n2\pi/Tt)$ have the period of T.

The calculation of a_n and b_n according to Equation 15.2 is called forward Fourier transformation and construction of the $y(t)$ via series of Equation 15.1 is called inverse Fourier transform.

The validity of the transformation can be shown by the use of the following relationship:

$$\frac{2}{T} \int_0^T \sin(n\omega_1 t) \cos(m\omega_1 t) dt = 0, \text{ for any integer } n \text{ and } m \qquad (15.4)$$

$$\frac{2}{T} \int_0^T \sin(n\omega_1 t) \sin(m\omega_1 t) dt = \delta_{nm},$$ (15.5)

$$\frac{2}{T} \int_0^T \cos(n\omega_1 t) \cos(m\omega_1 t) dt = \begin{cases} 2, & \text{for } n = m = 0 \\ \delta_{nm}, & \text{otherwise} \end{cases}$$ (15.6)

Note that $a_0/2$, according to Equation 15.2, is equal to

$$\frac{1}{2}a_0 = \frac{1}{2}\frac{2}{T}\int_0^T \cos(0\omega_1 t)y(t)dt = \frac{1}{T}\int_0^T y(t)dt = \overline{y(t)}$$ (15.7)

Thus, $a_0/2$ has a special meaning: it is the mean of the function over its period, that is, the base line, the DC offset, or the bias.

Also, a_n coefficients belong to cosines, thus they are responsible for the symmetrical part of the function (after offset removal). Consequently, b_n coefficients are in charge of the asymmetrical behavior.

Since each a_n or b_n coefficient corresponds to the oscillatory functions with the frequency $n\omega_1$, the set of a and b coefficients is often called *spectrum* when it is shown as the dependence on frequency.

15.1.1   Example: Fourier series for |t|

Let's find the Fourier series representation of the following periodic function:

$$y(t) = |t|, \quad -\pi < t < \pi$$

Since the function is symmetrical, we can be sure that all $b_n = 0$. The a_n coefficients are found by applying Equation 15.2:

$$\begin{cases} a_0 = \pi, \\ a_n = 0, & \text{for even } n \\ a_n = -\frac{4}{\pi n^2}, & \text{for odd } n \end{cases}$$

Their values are shown in Figure 15.2. As we can see, $a_0 = \pi$ is twice the mean of the $|t|$ on the $(-\pi, \pi)$ interval.

We can notice from Figure 15.2 that the a_n coefficients decrease very quickly as n grows, implying that the contribution of the higher n terms vanishes very quickly. Therefore, we can get quite a good approximation of $|t|$ with a truncated Fourier series, as shown in Figure 15.3.

This observation provides a basis for information compression. It is enough to know only the first 11 coefficients (by the way, half of them are zero) of the Fourier transform to reconstruct our function with minimal deviations from its true values at all possible times. If we need better precision, we can increase the number of the Fourier series coefficients.

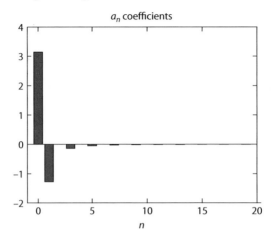

Figure 15.2   The first 20 a_n coefficients of the $|t|$ Fourier transform.

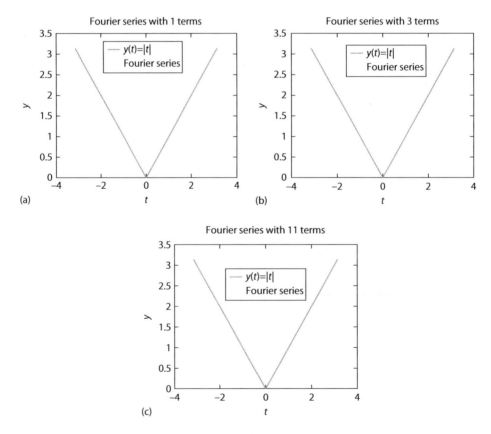

Figure 15.3   Approximation of the $|t|$ function by truncating the Fourier series at maximum $n = 1$ (a), $n = 3$ (c), and $n = 11$ (b).

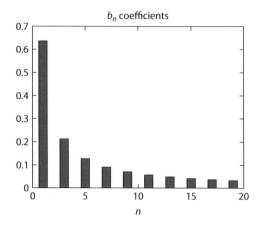

Figure 15.4   The first 20 b_n coefficients of the step function transform.

15.1.2 Example: Fourier series for the step function

Let's find the Fourier series for the step function defined as

$$\begin{cases} 0, & -\pi < x < 0, \\ 1, & 0 < x < \pi \end{cases}$$

This function is asymmetric (with respect to its mean value $1/2$), thus all $a_n = 0$ except the $a_0 = 1$. The b_n coefficients are

$$\begin{cases} b_n = 0, & n \text{ is even} \\ b_n = \frac{2}{\pi n}, & n \text{ is odd} \end{cases}$$

The values of the b_n coefficients are shown in Figure 15.4.

The b_n coefficients decrease at an inversely proportional rate to n. Therefore, we can hope that we can truncate the Fourier series and still get a good approximation of the step function. The result of truncation is shown in Figure 15.5. The coefficients do not drop as quickly as in the previous example; thus, we need more members in the Fourier series to get a good approximation. You may notice that at discontinuities where $t = -\pi, 0$, or π the approximation passes through midpoints $y = 1/2$ as we promised in Section 15.1. You may also notice a strange overshot near the discontinuities (ringing-like behavior). You may think this is the result of having a small number of members in the Fourier series, but it will not go away if we increase the number of expansion terms. This is known as the Gibbs phenomenon. Nevertheless, we have a good approximation of the function with a very small number of coefficients.

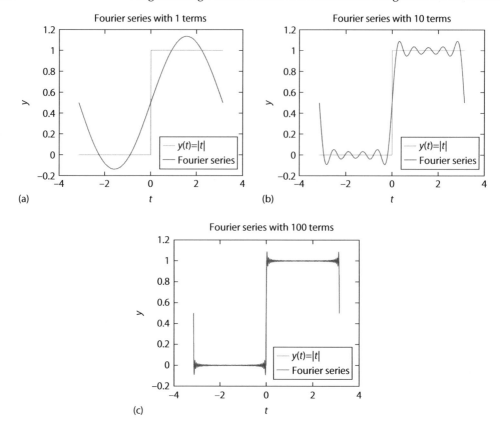

Figure 15.5 Approximation of the step function by truncating the Fourier series at maximum $n = 1$ (a), $n = 10$ (c), and $n = 100$ (b).

15.1.3 Complex Fourier series representation

Recall that

$$\exp(i\omega t) = \cos(\omega t) + i\sin(\omega t)$$

It can be shown that we can rewrite Equations (15.1) and (15.2) in the much more compact and symmetric notation without the eye-catching $1/2$ for the zero's coefficient:

Complex Fourier series

$$y(t) = \sum_{n=-\infty}^{\infty} c_n \exp(in\omega_1 t) \tag{15.8}$$

$$c_n = \frac{1}{T} \int_0^T y(t) \exp(-i\omega_1 n t) dt \tag{15.9}$$

The c_0 has the special meaning: the average of the function over the period or bias or offset.

The following connection exists between a_n, b_n, and c_n coefficients:

$$a_n = c_n + c_{-n} \tag{15.10}$$

$$b_n = i(c_n - c_{-n}) \tag{15.11}$$

You might ask yourself: what are those "negative" frequencies components for which c_{-n} is in charge? This does not look physical. It should not worry us too much, since $\cos(-\omega t) = \cos(\omega t)$ and $\sin(-\omega t) = -\sin(\omega t)$, that is, it is just a flip of the sign of the corresponding sine coefficient. We do this to make the forward and inverse transforms look alike.

15.1.4 Non-periodic functions

What to do if the function is not periodic? We need to pretend that it is periodic but on a very large interval, that is, $T \to \infty$. Under such an assumption, our discrete transform becomes a continuous one, that is, $c_n \to c_\omega$. In this case, we can approximate the sums in Equation (15.8) and (15.9) with integrals.[*]

Continuous Fourier transform

$$y(t) = \frac{1}{\sqrt{2\pi}} \int_{-\infty}^{\infty} c_\omega \exp(i\omega t)\, d\omega \tag{15.12}$$

$$c_\omega = \frac{1}{\sqrt{2\pi}} \int_{-\infty}^{\infty} y(t) \exp(-i\omega t)\, dt \tag{15.13}$$

Equations 15.12 and 15.13 require that $\int_{-\infty}^{\infty} y(t)\, dt$ exists, and it is finite.

Note that the choice of the normalization $1/\sqrt{2\pi}$ coefficient is somewhat arbitrary. In some literature, the forward transform has the overall coefficient of 1 and the inverse transform has the coefficient of $1/2\pi$. The opposite is also used. Physicists like the symmetric form shown previously in Equations (15.12) and (15.13).

15.2   Discrete Fourier Transform (DFT)

Our interest in the previous material is somewhat academic only. In real life, we cannot compute the infinite series, since it takes an infinite amount of time. Similarly, we cannot compute the exact integrals required for the forward Fourier transform, as doing so requires knowledge of the $y(t)$ at every point of time, which necessitates that we need an infinite amount of data to do the calculation. You might say that we have quite good methods to approximate the value of integrals covered in Chapter 9, but then we automatically limit ourself to a finite set of

---

[*] Here, we do the opposite to the rectangle integration method discussed in Section 9.2.

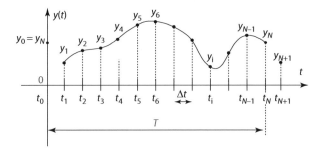

Figure 15.6   An example of discretely sampled signal $y(t)$ with period T.

points in time where we do the measurement of the $y(t)$ function. In this case, we do not need the infinite sum; it is enough to have only N coefficients of the Fourier transform to reconstruct N points of $y(t)$.

Assuming that $y(t)$ has a period T, we took N **equidistant** points such that spacing between them is $\Delta t = T/N$ (see Figure 15.6). The periodicity condition requires

$$y(t_{k+N}) = y(t_k) \tag{15.14}$$

We use the following notation $t_k = \Delta t \times k$ and $y_k = y(t_k)$ and define (see e.g., [9])

Discrete Fourier Transform (DFT)

$$y_k = \frac{1}{N} \sum_{n=0}^{N-1} c_n \exp\left(i\frac{2\pi(k-1)n}{N}\right), \text{ where } k = 1, 2, 3, \cdots, N \tag{15.15}$$

$$c_n = \sum_{k=1}^{N} y_k \exp\left(-i\frac{2\pi(k-1)n}{N}\right), \text{ where } n = 0, 1, 2, \cdots, N-1 \tag{15.16}$$

Notice that equations 15.15 and 15.16 do not have time in them at all! Strictly speaking, the DFT uniquely connects one periodic set of points with another; the rest is in the eye of the beholder. The notion of the spacing is required when we need to decide what is the corresponding frequency of the particular c_n: $f_1 \times n$, where $f_1 = T/N$ is the spacing (Δf) in the spectrum between two nearby c coefficients (see Figure 15.7). The other meaning of f_1 is the *resolution bandwidth* (RBW), that is, by construction, we cannot resolve any two frequencies with a spacing smaller than the RBW. The $f_s = 1/\Delta t$ is called the *sampling frequency* or the *acquisition frequency*. The Nyquist frequency $f_{Nq} = f_s/2 = f_1 N/2$ has a very important meaning that we will discuss later in Section 16.1.

Note the canonical placement of the normalization coefficient $1/N$ in Equation 15.15 instead of Equation 15.16. With this definition, c_0 is not the average

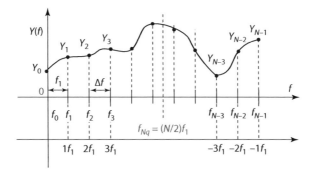

Figure 15.7   Sample spectrum: Fourier transform coefficient vs. frequency. Y_k is the same as c_k.

value of the function anymore; it is N times larger.* Unfortunately, pretty much every numerical library implements the DFT in this particular way, and MATLAB is not an exception.

There are several properties of the c_n coefficient; the proof of which is left as an exercise for the reader. The c coefficients are periodic:

$$c_{-n} = c_{N-n} \tag{15.17}$$

A careful reader would notice that the c_n and c_{-n} coefficients are responsible for the same absolute frequency $f_1 \times n$. Therefore, the spectrum is often plotted from $-N/2 \times f_1$ to $N/2 \times f_1$. It also has an additional benefit: if all y_k have no imaginary part, then

$$c_{-n} = c_n^* \tag{15.18}$$

that is, they have the same absolute value $|c_{-n}| = |c_n| = |c_{N-n}|$. This in turn means that for such real $y(t)$ the absolute values of the spectrum are symmetric either with respect to 0 or to the $N/2$ coefficient. Consequently, **the highest frequency** of the spectrum is **the Nyquist frequency** (f_{Nq}) and not the $(N-1) \times f_1$, which is $\approx f_s$ for the large N.

15.3 MATLAB's DFT Implementation and Fast Fourier Transform (FFT)

If someone implements Equation 15.16 directly, it would take N basic operations to calculate each c_n, and thus N^2 operations to do the full transform. This is extremely computationally taxing. Luckily, there is an algorithm, aptly named the *Fast Fourier Transform* (FFT) that does it in $\mathcal{O}(N \log N)$ steps [9], drastically speeding up the calculation time.

---

* This is what happens when mathematicians are in charge; they work with numbers and not with underlying physical parameters.

Matlab has built-in FFT realizations

- `fft(y)` for the forward Fourier transform

- `ifft(c)` for the inverse Fourier transform

Unfortunately (as we discussed in Section 15.2), MATLAB does FFT the same way as it is defined in Equation 15.16, that is, it does not normalize by N. So if you change number of points, the strength of the same spectral component will be different, which is unphysical. To get Fourier series coefficients (c_n) normalized, you need to calculate `fft(y)/N`. Nevertheless, the round trip normalization is maintained, that is, `y = ifft( fft(y) )`.

There is one more thing, which arises from the fact that MATLAB indexes arrays starting from 1. The array of the forward Fourier transform coefficients `c = fft(y)` has the shifted by 1 correspondence to c_n coefficients, that is, `c(n)`$= c_{n-1}$.

15.4   Compact Mathematical Notation for Fourier Transforms

The forward Fourier transform is often denoted as $\mathcal{F}$ and the coefficients of the forward transformation as $Y = (Y_0, Y_1, Y_2, \ldots, Y_{N-1}) = (c_0, c_1, c_2, \ldots, c_{N-1})$. In this notation, we refer to Y_n coefficients, instead of c_n coefficients. So, the spectrum of the time domain signal $y(t_k)$ is:

$$Y = \mathcal{F}y \tag{15.19}$$

The inverse Fourier transform is denoted as $\mathcal{F}^{-1}$:

$$y = \mathcal{F}^{-1}Y \tag{15.20}$$

15.5   DFT Example

Let's consider a very simple example that will help us to put together the previous material. We will sample and calculate the DFT for the following function:

$$y(t) = D + A_{one} \cos(2\pi f_{one}t) + A_{two} \cos(2\pi f_{two}t + \pi/4) \tag{15.21}$$

where

$D = -0.1$ is the displacement, offset, or bias of the function with respect to zero

$A_{one} = 0.7$ is the amplitude of the cosine with frequency $f_{one} = 10$ Hz

$A_{two} = 0.2$ is the amplitude of the $\pi/4$ shifted cosine with frequency $f_{two} = 30$ Hz.

For reasons that we explain later in Chapter 16, we chose the sampling frequency $f_s = 4f_{two}$. We run the following code to prepare time samples y_k for the time data set governed by Equation 15.21 and to calculate the corresponding DFT components $Y_n = $ `fft(y)`

Listing 15.1 `two_cos.m` (available at `http://physics.wm.edu/`
`programming_with_MATLAB_book/./ch_dft/code/two_cos.m`)

```
%% time dependence governing parameters
Displacement=-0.1;
f_one=10; A_one=.7;
f_two=30; A_two=.2;
f= @(t) Displacement + A_one*cos(2*pi*f_one*t) + A_two*cos
    (2*pi*f_two*t+pi/4);

%% time parameters
t_start=0;
T  = 5/f_one; % should be longer than the slowest
    component period
t_end = t_start + T;

% sampling frequency should be more than twice faster than
    the fastest component
f_s = f_two*4;
dt = 1/f_s; % spacing between sample points times
N=T/dt; % total number of sample points

t=linspace(t_start+dt,t_end, N); % sampling times
y=f(t); % function values in the sampled time

%% DFT via the Fast Fourier Transform algorithm
Y=fft(y);
Y_normalized = Y/N; % number of samples independent
    normalization
```

Let's first plot the time domain samples $y_k = y(t_k)$ and the underlying Equation 15.21.

Listing 15.2 `plot_two_cos_time_domain.m` (available at `http://physics.wm.`
`edu/programming_with_MATLAB_book/./ch_dft/code/`
`plot_two_cos_time_domain.m`)

```
two_cos;
%% this will be used to provide the guide for the user
t_envelope=linspace(t_start, t_end, 10*N);
y_envelope=f(t_envelope);
plot( t_envelope, y_envelope, 'k-', t, y, 'bo' );
fontSize=FontSizeSet; set(gca,'FontSize', fontSize );
xlabel('Time, S');
ylabel('y(t) and y(t_k)');
legend('y(t)', 'y(t_k)');
```

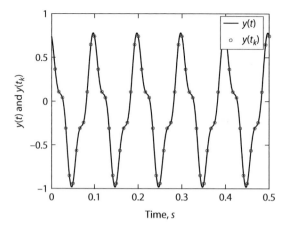

Figure 15.8   Sixty time domain samples and the underlying Equation 15.21.

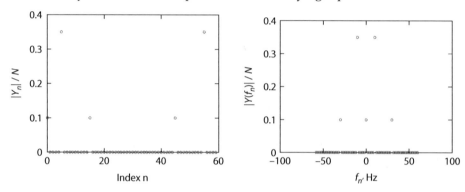

Figure 15.9   The normalized DFT coefficients for the time samples shown in Figure 15.8. The left panel depicts coefficient values versus their index (starting from 0). The right panel shows the spectrum, that is, coefficient values versus their corresponding frequency.

The result is shown in Figure 15.8. We can see five periods of the $y(t)$, though the function no longer resembles sine or cosine anymore due to combination of two cosines. Note that the $y(t)$ is shown as the guide to the eye only. The DFT algorithm has no access to it other than the 60 sampled points.

Let's now draw the $|Y_n|$ calculated by fft(y)

Listing 15.3   plot_two_cos_fft_domain.m (available at http://physics.wm. edu/programming_with_MATLAB_book/./ch_dft/code/ plot_two_cos_fft_domain.m)

```
two_cos;

n=(1:N) - 1; % shifting n from MATLAB to the DFT notation

plot( n, abs(Y_normalized), 'bo' );
fontSize=FontSizeSet; set(gca,'FontSize', fontSize );
xlabel('Index n');
ylabel('|Y_n| / N');
```

The result is shown in the left panel of Figure 15.9. Note that we normalized the result of the DFT by number of points $N = 60$. This allows us to see the true nature of the Fourier transform coefficients: recall (see Section 15.2) that the Y_0 coefficient normalized by N corresponds to the mean or displacement of the function, which is -0.1 as we set it for Equation 15.21. We might wonder why there are four more non-zero coefficients if we have only two cosines with two distinct frequencies. This is due to reflection property of the DFT, that is, Y_{-n} and Y_{N-n} corresponds to the same frequency. This is better shown if we plot the spectrum, that is, Y_n versus the corresponding frequency. This is done with

Listing 15.4 `plot_two_cos_freq_domain.m` (available at `http://physics.wm.edu/programming_with_MATLAB_book/./ch_dft/code/plot_two_cos_freq_domain.m`)

```
two_cos;

freq = fourier_frequencies(f_s, N); % Y(i) has frequency
    freq(i)

plot( freq, abs(Y_normalized), 'bo' );
fontSize=FontSizeSet; set(gca,'FontSize', fontSize );
xlabel('f_n, Hz');
ylabel('|Y(f_n)| / N');
```

the Y_n index transformation to the frequency is done with the helper function

Listing 15.5 `fourier_frequencies.m` (available at `http://physics.wm.edu/programming_with_MATLAB_book/./ch_dft/code/fourier_frequencies.m`)

```
function spectrum_freq=fourier_frequencies(SampleRate, N)
        %% return column vector of positive and negative
            frequencies for DFT
        % SampleRate - acquisition rate in Hz
        % N - number of data points

        f1=SampleRate/N; % spacing or RBW frequency

        % assignment of frequency,
        % recall that spectrum_freq(1) is zero frequency,
            i.e. DC component
        spectrum_freq=(((1:N)-1)*f1).'; % column vector

        NyquistFreq= (N/2)*f1; % index of Nyquist
            frequency i.e. reflection point

        %let's take reflection into account
        spectrum_freq(spectrum_freq>NyquistFreq) =-N*f1+
            spectrum_freq(spectrum_freq>NyquistFreq);
end
```

Now we have the canonical spectrum shown in the right panel of Figure 15.9. Notice that the spectrum of the absolute values of Y_n is fully symmetric, that is, mirrored around $f = 0$, as it predicted by Equation 15.18. Now, we see that the spectrum has only two strong frequency components at 10 Hz and 30 Hz. This is in complete accordance with Equation 15.21 and the values of $f_{one} = 10$ Hz and $f_{two} = 30$ Hz. Now, let's examine the components' values: $Y(10)$ Hz $= 0.35$, which is exactly half of the $A_{one} = 0.70$. A similar story is seen for the other frequency component $Y(30$ Hz$) = 0.1$. This is due to Equations (Equation 15.10) and (15.11) and the fact that $y(t)$ has no imaginary part. Note that Y_n themselves can be complex even in this case.

15.6   Self-Study

Problem 15.1
Have a look at the particular realization of the N point forward DFT with the omitted normalization coefficient:

$$C_n = \sum_{k=1}^{N} y_k \exp(-i2\pi(k-1)n/N)$$

Analytically prove that the forward discrete Fourier transform is periodic, that is, $c_{n+N} = c_n$. Note: recall that $\exp(\pm i2\pi) = 1$.
Does this also prove that $c_{-n} = c_{N-n}$?

Problem 15.2
Use the proof for the previous problem's relationships and show that the following relationship holds for any sample set that has only real values (that is, no complex part)

$$c_n = c_{N-n}^*$$

where:
   * depicts the complex conjugation.

Problem 15.3
Load the data from the file `'data_for_dft.dat'`.* provided at the book's web page. It contains a table with y vs t data points (the first column holds the time, the second holds y). These data points are taken with the same sampling rate.

1. What is the sampling rate?
2. Calculate forward DFT of the data (use MATLAB built-ins) and find which two frequency components of the spectrum (measured in Hz not rad^{-1}) are

---

* The file is available at http://physics.wm.edu/programming_with_MATLAB_book/./ ch_dft/data/data_for_dft.dat

the largest. Note: I refer to the real frequency of the sin or cos component, that is, only positive frequencies.

3. What is the largest frequency (in Hz) in this data set that we can scientifically discuss?

4. What is the lowest frequency (in Hz) in this data set that we can scientifically discuss?

CHAPTER 16

Digital Filters

This chapter focuses on the discussion of the discrete Fourier transform in its application to digital filters. We discuss Nyquist's criteria for capturing or digitization of continuous signals, showing examples of simple digital filters and discussing the artifacts that arise during digital filtering.

One of the main applications for the digital Fourier transform (DFT) is digital filtering, that is, reducing unwanted frequency components or boosting the desired ones. We all have seen it in the form of an equalizer in a music player, which allows one to adjust the loudness of low-frequency components (bass) relative to the middle- and the high-frequency components (treble) of the sound spectrum. It used to be done with analog electronics components filters, but these days, with the proliferation of micro controllers, it is often done digitally via the DFT.

16.1 Nyquist Frequency and the Minimal Sampling Rate

Before we can filter any data, we should acquire it. The main question is what the sampling rate ($f_s = 1/\Delta t$) of the data acquisition should be.

Recall the discussion in Section 15.2, where we showed that the highest observable frequency in the DFT spectrum is $\approx f_s/2$. This brings us to:

> **Nyquist–Shannon sampling criteria**
>
> If our signal has the highest frequency f_{max} then we need to sample it with
>
> $$f_s > 2f_{max} \tag{16.1}$$
>
> Pay attention to the $>$ relationship!

The Nyquist–Shannon sampling criteria is quite often used in reverse form: one cannot acquire frequencies in the signal above the Nyquist frequency $f_{Nq} = f_s/2$.

This criteria is not very constructive. How would we know what the highest frequency of the signal is? Sometimes we know it from the physical limitations of our apparatus. If we don't know the highest frequency, we should sample the signal with some sampling frequency. If at the high-frequency end of the spectrum, the strength of components drops to zero, then we have sampled fast enough, so

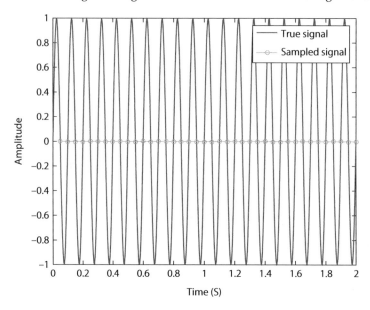

Figure 16.1   The signal described by Equation 16.2 (lines) and the undersampled signal acquired with $f_s = 2f_{signal}$ (circles and lines).

we might even try to reduce the sampling frequency. Otherwise, we are *under-sampling*. We must increase the sampling frequency until we see the high end of the spectrum asymptotically small.

Words of wisdom

Choosing the sampling frequency is the most important part of the data acquisition. No amount of post-processing will be able to recover or restore a signal that was acquired with a wrong sampling frequency.

16.1.1   Undersampling and aliasing

We will see how an undersampled signal might look in the examples shown in this section. This section will also sample the signal described by the following equation:

$$y(t) = \sin(2\pi 10 t) \tag{16.2}$$

that is, this is the sinusoidal signal with the signal frequency $f_{signal} = 10$ Hz.

At first, we sample our signal with $f_s = 2f_{signal}$. As we can see in Figure 16.1, the sampled signal appears as a straight line, even though it is evident that the underlying signal is sinusoidal. We clearly used the wrong sampling frequency. This example emphasizes the $>$ relationship in Equation 16.1. Note that the lines

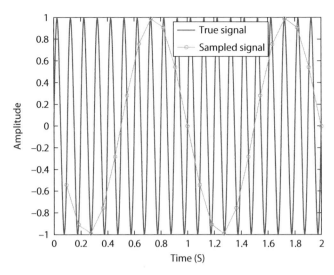

Figure 16.2   The signal described by Equation 16.2 (lines) and the undersampled signal acquired with $f_s = 1.1 f_{signal}$ (circles and lines).

connecting the sampling points are just to guide an eye. The DFT has no notion of the signal values in between the sampled points.

In the following example, we sample with $f_s = 1.1 f_{signal}$, that is, our sampling frequency does not satisfy the Nyquist–Shannon criteria. As we can see in Figure 16.2, the sampled signal does not reproduce the underlying signal. Moreover, the undersampled signal appears as a signal with a lower frequency. This phenomenon is called *aliasing* or *ghosting*. It arises due to the periodicity of the DFT spectrum, as high-frequency components require the existence of the c_M components, where $M > N$ but, we recall that $C_{-n} = C_{N-n}$; see Equation 15.17. Thus $c_M = c_m$ where $m = M - N \times l$ and l is an integer. In other words, if the signal is undersampled, that is, the sampling frequency does not satisfy the previous Section 16.1, and a high-frequency component appears as a low-frequency one. Consequently, in the undersampled spectrum, we see the ghost frequency components:

$$f_{ghost} = |f_{signal} - l \times f_s| \tag{16.3}$$

For the case depicted in Figure 16.2, we see the appearance of the signal with $f_{ghost} = 0.1$ Hz or period of 1 s.[*]

In some cases, the aliasing makes even stranger-looking sampled signals that do not resemble the underlying signal at all. This is illustrated in Figure 16.3, where the $f_s = 1.93 f_{signal}$.

---

[*] If you ever use a digital oscilloscope, be aware of the aliasing phenomenon. If you choose your acquisition rate wrongly, you will see nonexistent signals.

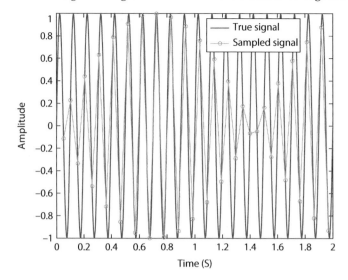

Figure 16.3   The signal described by Equation 16.2 (lines) and the undersampled signal acquired with $f_s = 1.93 f_{signal}$ (circles and lines).

> **Words of wisdom**
>
> If you cannot sample fast enough, build a low-pass electronics filter that removes fast frequency components larger than obtainable $f_s/2$. Otherwise, the digitized signal will be contaminated by ghost signals.

16.2   DFT Filters

The material outlined in the previous section about the importance of the proper sampling frequency is strictly speaking outside the scope of this book. It is in the domain of data acquisition and instrumental science. Nevertheless, it is always a good idea to apply sanity checks to the data and see what could have gone wrong before you start analyzing the data.

For now, we assume that someone gave us the data and we have to work with what we received. As we discussed at the beginning of the chapter, the job of a digital filter is to somehow modify the spectrum of the signal, that is, to boost or suppress certain frequencies, and then reconstruct the filtered signal.

The recipe is the following:

- Calculate the DFT (use MATLAB's `fft`) of the signal

- Have a look at the spectrum and decide which frequencies are to be modified

- Apply a filter that adjusts the amplitudes of the frequencies we are interested in

- For signals belonging to the real numbers domain: if you attenuate the component with the frequency f by g_f, you need to attenuate the component at $-f$ by g_f^*. Otherwise, the inverse Fourier transform, that is, the reconstructed signal, will have a non-zero imaginary part

- Calculate inverse DFT (use MATLAB's `ifft`) of the filtered spectrum

- Repeat if needed

Mathematical representation of the digital filtering

$$y_{filtered}(t) = \mathcal{F}^{-1}\left[\mathcal{F}(y(t)) \times G(f)\right] = \mathcal{F}^{-1}\left[Y(f) \times G(f)\right] \qquad (16.4)$$

where

$$G(f) = Y_{filtered}(f)/Y(f) \qquad (16.5)$$

is the frequency-dependent *gain* function.* The $G(f)$ controls how much we change the corresponding spectral component.

The Equation 16.4 looks quite intimidating, but we will show in Section 16.2.1 that the filtering is very easy. Simultaneously, we will learn some standard filters and their descriptions.

16.2.1 Low-pass filter

In all following examples, we will work with the following signal:

$$y(t) = 10\sin(2\pi f_1 t) + 2\sin(2\pi f_2 t) + 1\sin(2\pi f_3 t) \qquad (16.6)$$

where:

$f_1 = 0.02$ Hz is the slow component
$f_2 = 10f_1$ Hz
$f_3 = 0.9$ Hz is the fast component.

This signal is depicted in Figure 16.4a. The spectrum of this signal sampled with $f_s = 2$ Hz on the interval of 100 s, is shown in Figure 16.4b. As expected, it consists of three strong frequency components corresponding to f_1, f_2, and f_3.

The *low-pass filter* is a filter that strongly suppresses or attenuates high-frequency components of the spectrum, while keeping low-frequency components mostly intact. We specifically focus on the brick wall low-pass filter described by the following gain equation:

$$G(f) = \begin{cases} 1, |f| \leq f_{cutoff} \\ 0, |f| > f_{cutoff} \end{cases} \qquad (16.7)$$

---

* In spite of the name, it is quite often less than unity.

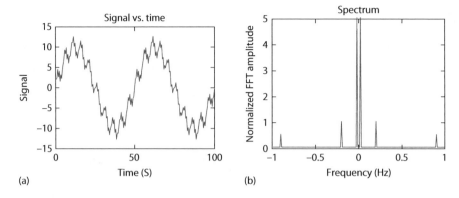

Figure 16.4   The signal described by Equation 16.6 (a) and its spectrum (b).

The gain function is usually complex, so it is often shown in the Bode plot representation where we plot the absolute value (magnitude) of the gain versus frequency in the upper sub-plot and the complex angle (or phase) of the gain versus frequency in the lower sub-plot. For this filter, it is shown in Figure 16.5a. The name "brick wall" comes from the very sharp transitions of the filter gain near the cutoff frequency f_{cutoff}, which is equal to 0.24 Hz in this case.

We obtain the filtered y signal with the following code

```
freq=fourier_frequencies(SampleRate, N);
G=ones(N,1); G( abs(freq) > Fcutoff, 1)= 0;
y_filtered  =  ifft(  fft( y ) .* G  )
```

As you can see, it is very simple. The filter strength is calculated and assigned at the second line of previous code, and the filter application is done at the last line. The indispensable function fourier_frequencies connects a particular index in the DFT spectrum to its frequency (we discussed its Listing 15.5 in Section 15.5).

The filtered spectrum is shown in Figure 16.5b. As expected, the spectral component with the frequency f_3 is now zero, since it lies beyond the cutoff frequency. We now have only f_1 and f_2 in the spectrum. Consequently, the filtered signal does not have a high-frequency (f_3) component, as is shown in the comparison of the filtered and raw signals in Figure 16.5c. The filtered signal is much smoother, that is, missing the high-frequency components. Thus, application of the low-pass filter is sometimes referred to as *smoothing* or *denoising*.[*]

16.2.2   High-pass filter

The *high-pass filter* is the exact opposite of the low-pass filter, that is, it attenuates low-frequency components and leaves high-frequency components intact. For

---

[*] "Denoising" is actually a misnomer since the useful signal can be located at high frequencies as well.

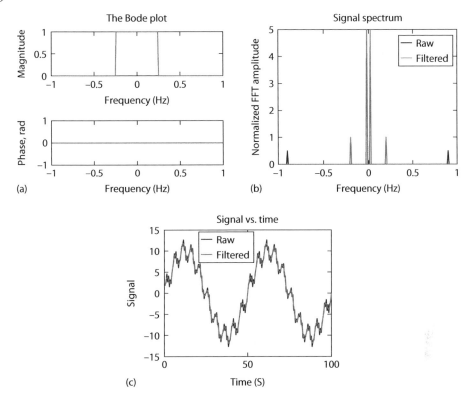

Figure 16.5   Bode plot of the brick wall low-pass filter (a). Comparison of the filtered and unfiltered spectra (b) and signals (c).

example, the brick wall high-pass filter can be describe by the following equation:

$$G(f) = \begin{cases} 0, |f| \le f_{cutoff} \\ 1, |f| > f_{cutoff} \end{cases} \tag{16.8}$$

The Bode plot of this filter is shown in Figure 16.6a. The MATLAB implementation of the brick wall high-pass filter is shown in the following code:

```
freq=fourier_frequencies(SampleRate, N);
G=ones(N,1); G( abs(freq) < Fcutoff, 1)= 0;
y_filtered   =   ifft(  fft( y ) .* G  )
```

The filtered spectrum, missing the low-frequency component f_1, is shown in Figure 16.6b. As we can see in Figure 16.6c, the high-pass filter gets rid of the slow envelope from the raw signal.

16.2.3   Band-pass and band-stop filters

The *band-pass filter* allows only frequency components within a certain bandwidth (f_{bw}) in the vicinity of the specified central frequency (f_c) to pass, while all other

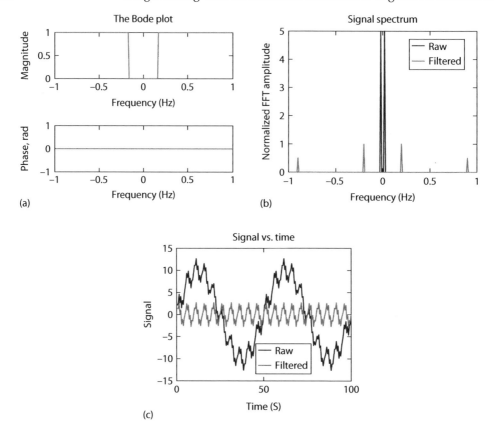

Figure 16.6   Bode plot of the brick wall high-pass filter (a). Comparison of the filtered and unfiltered spectra (b) and signals (c).

frequencies are strongly attenuated. For example, the brick wall band-pass filter is described by the following equation:

$$G(f) = \begin{cases} 1, ||f| - f_c| \le \frac{f_{bw}}{2} \\ 0, ||f| - f_c| > \frac{f_{bw}}{2} \end{cases} \tag{16.9}$$

It can be implemented by the following MATLAB code:

```
freq=fourier_frequencies(SampleRate, N);
G=ones(N,1); G( abs(abs(freq)-Fcenter) > BW/2, 1)=0;
y_filtered   =   ifft(   fft( y ) .* G  )
```

The *band-stop* (or *band-cut*) filter is the exact opposite of the band-pass filter. The brick wall band-stop filter is described by the following equation:

$$G(f) = \begin{cases} 0, ||f| - f_c| \le \frac{f_{bw}}{2} \\ 1, ||f| - f_c| > \frac{f_{bw}}{2} \end{cases} \tag{16.10}$$

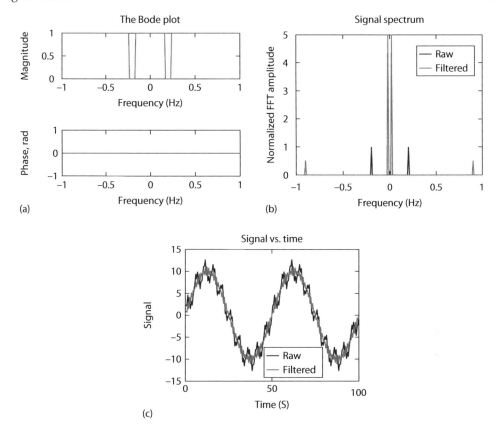

Figure 16.7 Bode plot of the brick wall band-stop filter (a). Comparison of the filtered and unfiltered spectra (b) and signals (c).

The MATLAB implementation is by the following code

```
freq=fourier_frequencies(SampleRate, N);
G=zeros(N,1); G( abs(abs(freq)-Fcenter) > BW/2, 1)=1;
y_filtered  =  ifft(  fft( y )  .* G  )
```

The Bode plot of the filter with $f_c = f_2 = 0.2$ Hz and $f_{bw} = 0.08$ Hz is shown in Figure 16.7a. The band-stop filter with these parameters will remove f_2 from our signal spectrum, as shown in Figure 16.7b. The filtered signal now looks like slow envelope with the frequency f_1 with the high-frequency (f_3) "fuss" on top of it, as shown in Figure 16.7c.

16.3 Filter's Artifacts

The ease of the implementation of the brick wall filters comes at a price: they often produce ring down artifacts that were not present in the original raw signal.

Let's have a look at the signal and its spectrum depicted Figure 16.8a and b, respectively. When we apply the brick wall low-pass filter with $f_{cutoff} = 0.24$ Hz

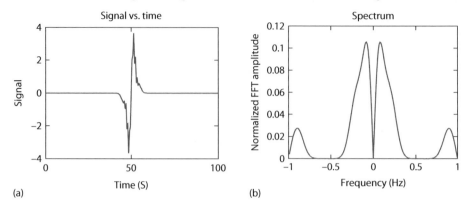

Figure 16.8   The sample signal (a) and its spectrum (b).

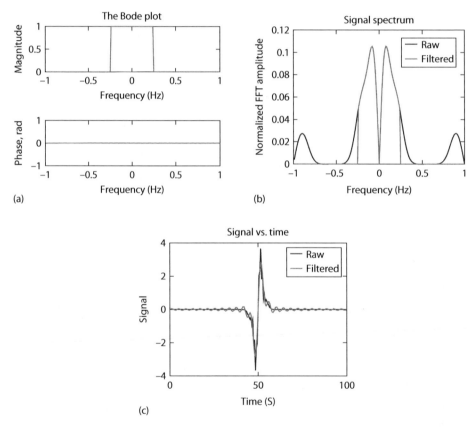

Figure 16.9   Bode plot of the brick wall low-pass filter (a). Comparison of the filtered and unfiltered spectra (b) and signals (c).

(see its Bode plot in Figure 16.9a), we obtain the filtered spectrum shown in Figure 16.9b. The significant discontinuity at 0.24 Hz produces a large ring-down-like disturbance on the filtered signal, as we can see in Figure 16.9c. The easiest way to avoid this is to use a filter without discontinuities in the spectrum. For

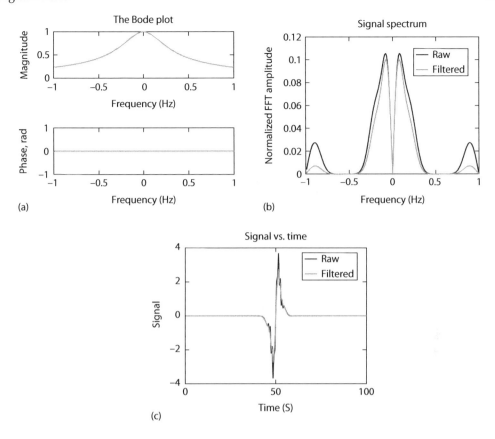

Figure 16.10 Bode plot of the smooth low-pass filter (a). Comparison of the filtered and unfiltered spectra (b) and signals (c).

example, we can construct the smooth low-pass gain function according to the following equation:

$$G(f) = \left| \frac{1}{1 + i(f / f_{\text{cutoff}})} \right| \qquad (16.11)$$

with the Bode plot depicted in Figure 16.10a.

This filter is weaker than its brick wall counterpart. Thus, the high-frequency components are not as strongly suppressed (see Figure 16.10b). Nevertheless, we successfully removed high frequencies from the raw signal without introducing artifacts (see Figure 16.10c).

16.4 Windowing Artifacts

The time domain discontinuities generate spurious frequency components in the DFT spectrum. Such discontinuities are often located at the beginning and the end of the acquisition period since this period is often arbitrarily chosen.

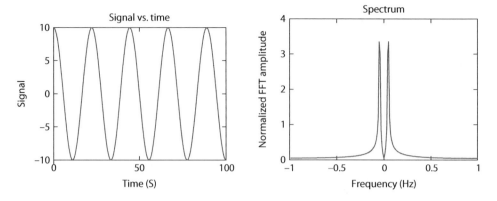

Figure 16.11 The cosine signal acquired for the time not matching it own period and its DFT spectrum.

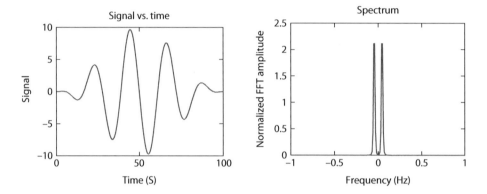

Figure 16.12 Signal and its spectrum after application of the Hann window.

For example, even if we sample a pure cosine signal, the ends of the acquired signals might not match, as shown in Figure 16.11a. We expect a spectrum with the single frequency matching the cosine one, but discontinuities generate nonexistent frequency components, as shown in Figure 16.11b. For example, we see non-zero spectral components beyond 0.5 Hz frequency for the underlying cosine with 0.045 Hz frequency.

To avoid this, usually some sort of *windowing* function ($w(t)$) is applied and then the DFT calculated on $y(t) \times w(t)$. There are many window functions,* but they all have a common property: they asymptotically approach zero at the beginning and the end of the acquisition time to remove discontinuity at the ends. For example, the Hann window coefficients are given by the following equation:

$$w_n = \frac{1}{2}\left[1 - \cos\left(2\pi\frac{n-1}{N-1}\right)\right] \tag{16.12}$$

---

* Some of the most popular are Hamming, Tukey, Cosine, Lanczos, Triangulars, Gaussians, Bartlett–Hann, Blackmans, Kaisers.

Our cosine signal with applied Hann's window looks like the signal shown in Figure 16.12a. The resulting DFT spectrum shape better matches the single frequency spectrum of the original cosine, as we can see in Figure 16.12b.

We are calculating fft(y.*w), that is, the spectrum of the modified function, so the DFT spectrum strength and shape should not exactly match the underlying signal. Nevertheless, a window function often drastically improves the fidelity of the spectrum, although it often reduces the spectral RBW, which now is $\sim 1/T_{window} < T_{acq}$; here, T_{window} is characteristic time where window function is large, and T_{acq} is the full acquisition time.

16.5   Self-Study

Problem 16.1

Download the wave file 'voice_record_with_a_tone.wav'.* It is the audio file. If you play it, you will hear a very loud single-frequency tone, which masks the recorded voice. Apply an appropriate band-stop filter to hear the message. What is the message?

To obtain audio data, use the following commands. This assumes that the audio file is in the current folder of MATLAB.

```
[ydata, SampleRate]=audioread('voice_record_with_a_tone.
    wav', 'double');
% the following is needed if you want to save the filtered
    signal
info=audioinfo('voice_record_with_a_tone.wav');
NbitsPerSample=info.BitsPerSample;
```

After execution of this code, the ydata variable will hold the amplitudes of audio signal sampled equidistantly with the sampling rate stored in the SampleRate variable. Note that columns in ydata correspond to audio channels. So, there could be more than one. However, it is sufficient to process only one channel for this problem.

Once you have filtered your data, it is a good idea to normalize it to 1. This will make it louder.

You can play audio data within MATLAB with the following command:

```
sound(y_filtered(:,1), SampleRate);
```

Alternatively, you can save it into a wave audio file with the following command:

```
audiowrite('voice_record_filtered.wav', y_filtered,
    SampleRate, ...
  'BitsPerSample', NbitsPerSample);
```

---

* The file is available at http://physics.wm.edu/programming_with_MATLAB_book/./
  ch_functions_and_scripts/data/voice_record_with_a_tone.wav

References

[1] University of South Florida. Holistic numerical methods. `http://mathforcollege.com/nm/`. Accessed: 2016-11-09.

[2] National Institute of Standards and Technology. A statistical test suite for the validation of random number generators and pseudo random number generators for cryptographic applications, 2010. `http://csrc.nist.gov/groups/ST/toolkit/rng/documentation_software.html`. Accessed: 2016-10-09.

[3] R. Bellman. Dynamic programming treatment of the travelling salesman problem. *Journal of the ACM*, 9(1):61–63, 1962.

[4] P. R. Bevington. *Data Reduction and Error Analysis for the Physical Sciences*. New York, McGraw-Hill, 1969.

[5] C. Darwin. *On the Origin of Species by Means of Natural Selection, or the Preservation of Favoured Races in the Struggle for Life*, 1st edn. London, John Murray, 1859.

[6] V. Granville, M. Krivanek, and J. P. Rasson. Simulated annealing: A proof of convergence. *IEEE Transactions on Pattern Analysis and Machine Intelligence*, 16(6):652–656, 1994.

[7] D. E. Knuth. *The Art of Computer Programming, Volume 4 A: Combinatorial Algorithms, Part 1*, 3rd edn. Boston, Addison-Wesley Professional, 2011.

[8] N. Metropolis, A. W. Rosenbluth, M. N. Rosenbluth, A. H. Teller, and E. Teller. Equation of state calculations by fast computing machines. *Journal of Physical Chemistry,*, 21:1087–1092, 1953.

[9] W. H. Press, S. A. Teukolsky, W. T. Vetterling, and B. P. Flannery. *Numerical Recipes 3rd Edition: The Art of Scientific Computing*, 3rd edn. New York, Cambridge University Press, 2007.

[10] C. Ridders. A new algorithm for computing a single root of a real continuous function. IEEE Transactions on *Circuits and Systems*, 26(11):979–980, 1979.

Index